ALL MAPPED OUT

ALL MAPPED OUT

HOW MAPS SHAPE US

MIKE DUGGAN

REAKTION BOOKS

For Elliot

Published by Reaktion Books Ltd
Unit 32, Waterside
44–48 Wharf Road
London N1 7UX, UK
www.reaktionbooks.co.uk

First published 2024
Copyright © Mike Duggan 2024

Printed and bound in Great Britain by Bell & Bain, Glasgow

A catalogue record for this book is available from the British Library

ISBN 978 1 78914 836 7

CONTENTS

Bedolina petroglyph pictured in the
Seradina-Bedolina Municipal Archaeological Park, Italy, 2008.

Introduction

C arved out of the side of the Val Camonica valley in northern Italy more than 15,000 years ago, the Bedolina petroglyph is thought to be one of the earliest known topographical maps – a map showing the shape and physical characteristics of a place. Despite the interest the map has provoked from the academic community and the local tourist board, the jury is still out on what it really shows. It appears to represent an ancient human settlement with paths, buildings, farming plots, streams and wells, among other features. A tracing of the map makes these elements stand out even more. Here we can see what looks like a complex settlement that accommodates different activities and social arrangements. Some argue that the etchings highlight meeting places, trading zones, farms and homes, separated by paths, rivers or social divisions. Others are less sure, suggesting that such interpretation is based on our understanding of maps today. They argue that it is important to read the Bedolina Map as a product of its time, when symbolic representation had a different meaning.

The map historian Matthew Edney urges caution when equating ancient map-like images with the maps that are used today.[1] People have been representing their surroundings through imagery for millennia, but it must not be forgotten that images, and especially maps, have a long history of development entangled with the norms and customs of different societies, each with its own idea about what counts as a map. Edney argues that the history of cartography is all too often mistakenly thought to have followed a precise chronology, from Claudius Ptolemy's *Geographia* in the second century to al-Sharīf al-Idrīsī's *Book of Roger* in the twelfth, and from the 'Padrón Real' of Spain's Casa de Contratación in the sixteenth century to Giovanni Domenico Cassini's 'Mappe Monde' at the end of the seventeenth. If we consider the history of map-making and all that has changed in the development of cartography as a science, we might pause and rethink our assumption that the Bedolina Map and maps of today are broadly the same. We could, for example, consider that the Bedolina Map was made long before the widespread adoption of the god's-eye view we are now used to (a view made possible only after the invention of aeronautical sensing technology), before the widespread use of map scales and keys, and before the standardized latitude/longitude coordinates system that we associate with maps.[2] These standards, which have become ingrained into our way of reading maps, were developed thousands of years later, as the science of cartography emerged in eighteenth-century Europe. And yet, they have shaped our understanding of maps, throughout history, ever since. We might also consider that the Bedolina Map – if it is indeed a map – was carved into the rock by a Neolithic community living in a completely different culture from our own, with different societal structures.

Cartography is assumed to reference mapmaking of any kind, but the term emerged only relatively recently, in mid-nineteenth-century French, from the coupling of *carte* (map, card) and *graphie* (writing).[3] It grew in popularity alongside the professionalization of mapmaking and its establishment as a science. Edney has written that a 'cartographic ideal' emerged around the same time. This was an ideal based on the style, form and conventions of these maps, which are recognized today as the characteristics of maps: among them the aerial perspective, universal coordinate systems, legend, place pins and contour lines. Edney argues that this ideal has

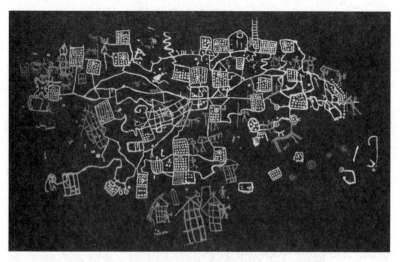

Archaeological tracing of the Bedolina petroglyph.

gripped society's view of maps ever since, to the extent that all maps, regardless of their context of production or use throughout history, have been mistakenly understood through this lens.[4]

When we consider that most of us learned about maps with cartographic standards already in place, most likely in early years education during geography lessons or orienteering exercises, our points of comparison with the Bedolina Map become misaligned. The paths, plots, buildings and streams that we assume to be represented on that map might be symbolic of something else entirely; perhaps they are used to represent storage plots, cattle pens, imprisoned people or spiritual iconography. Without a scale to think with, we also have no idea how close together or far apart the symbols on the map are supposed to be. It might not even be a representation related to the geography of the valley. Just because the Bedolina Map was found overlooking the valley does not mean that it is a representation of what is found on the valley floor.

Setting aside these provocations, we should also recognize that scholars have yet to come to a consensus on what the Bedolina Map truly shows. Archaeologists claim that it accurately represents agricultural plots from the Bronze Age valley floor below. Map historians, on the other hand, have argued that the map is in a mountainous, rocky, hardly arable area, a circumstance that challenges the notion that it is a map of systematic farming. Moreover, they argue that without any key on the map, or any local archaeological finds to indicate what the symbols on the map might mean, we have no real idea what it represents. All we know is that it fits with our idea of a geographical map: a two-dimensional image of Earth seen from above. The Italian geographer and cartographic theorist Emanuela Casti concludes that 'ultimately, all the hypotheses advanced so far lack hard evidence or solid arguments: the Bedolina map has yet to be thoroughly deciphered. We are groping in the dark.'[5] This follows Catherine Delano-Smith's study of prehistoric maps in Europe, in which she lays out and compares different cartographic styles dated to this period. We are reminded not to assume that ancient map-like representations are topographic.[6] On Bedolina and Giadighe, another petroglyph in the Val Camonica region, she writes that 'it remains doubtful that even these two examples can in fact be seen as marking the introduction of the use of maps as factual records in the prehistoric era, although such a transition had taken place by early historical times.'[7]

Even when considering maps that are made and used today, most people take for granted what a map is and what it is for. This is clearly highlighted in non-Western and Indigenous mapping cultures, where cartography as those in the West know it is not the be-all and end-all of spatial representation. For example, First Nations peoples of Australia have traditions of using song and ceremony to map and keep alive their history with the landscape.[8] Song-maps (sometimes named 'songlines', 'dreaming tracks' or 'strings') show that not all geographical maps are two-dimensional or graphical. They can be vocal and deeply embedded in cultural life. Many Indigenous communities have their own songlines that map out the story of Earth and sky, as well as their relations to them.

These songs are akin to folk tales, and must be sung in ceremony, in a specific order from memory, by learned members of the community. They sing of cultural laws, ancestorial origin stories, myths and legends, trading routes, celestial formations, local and distant landmarks and how these came to be significant to their culture, from natural phenomena with spiritual associations to historical locations where significant social events have taken place, and to the living creatures alongside which they live and have lived. Today there are as many songlines as there are ancient pathways across the Australian outback, as well as those now buried (but not forgotten) under modern societies. As songlines are sung at local and regional gatherings, the relationship Indigenous peoples have with the Earth – where land, water, wind, animals and body are inextricably intertwined – is cemented and passed on to others. In doing so, a metaphorical but still navigationally useful map emerges. By singing about the Earth and their place within it, these peoples have developed not only a spatial understanding of where things are, but a comprehension of why different places are significant to their ways of life.

Indigenous Pacific Islanders have a history of using environmental cues to navigate open water. Looking to sky and sea for direction, experienced Polynesian navigators successfully sailed between islands in the Pacific without maps for centuries before colonizer-explorers from Europe arrived. This skill was developed over time, not through the learning of nautical charts and the use of navigational instruments, but through practical experience and intergenerational storytelling, whereby geographical knowledge of the sun, stars, winds, currents, swells, depths, rocks, corals, birds and fish schools was passed on.[9]

Navigation in this context is so much more than a technical skill; it is about finding one's way *with* the Earth rather than becoming a master *of* it, which tends to be what Western navigational skills are geared towards. Some argue that navigation and wayfinding are not even the same thing in this context.[10] The former has developed into the scientific practice of moving across the Earth in the most efficient way, using Western standards of abstract cartography, systems of coordinates and technological instruments. The latter is a more deeply embedded and embodied cultural practice central to Indigenous lives and their relationship to their environment. In this understanding, there is a weight to wayfinding that cannot be conveyed in the same way as navigational instructions, by passing on directions or charting coordinates. Fetaui Iosefo, Stacy Holman Jones and Anne Harris, researchers into the knowledge of Indigenous Pacific peoples, write that 'at the heart of wayfinding in the Pacific is the ancestral genealogical connection to the environment and its people.'[11]

It is important to note that people who are not Indigenous Pacific Islanders, myself included, cannot claim to be able to convey the cultural significance of Pacific wayfinding. To do so would risk repeating the mistakes of anthropological writings, in which the written practice of translation has often failed to convey the lived reality of another culture. Wayfinding in this context consists of knowledge systems that exist beyond the written word, that must be accumulated through practice over time. They demand respect. Those without permission cannot simply speak about them with authority just because they have an interest in maps.[12] What they can do, however, is learn to listen to, rather than dismiss, Indigenous peoples who practise wayfinding in this way.[13]

Wayfinding remains a cultural practice of journeying through life for many Indigenous Pacific peoples, but traditional navigational techniques have been in danger of being lost forever in some places owing to the systematic erasure of local knowledge by colonial rule, for example through imposed educational and religious systems. In response, there has in the last few decades been a resurgence in people trying to keep the traditions of Indigenous sailing alive. Most notably, this began in 1976 with the first ocean voyage of *Hōkūlea*, a Polynesian double-hulled canoe built by the Polynesian

Voyaging Society (PVS, established in 1975). Led by the famous Micronesian wayfinding master Pius 'Mau' Piailug, *Hōkūlea* made the journey from Hawaii to Tahiti in 31 days without any of the navigational equipment commonly used at the time. This – said to be the first voyage using traditional techniques to leave Hawaii in six hundred years – was a remarkable achievement considering how reliant seafarers had become on so-called scientific navigational technology to sail across the ocean, even before the widespread use of GPS on sailing vessels today.[14]

Mau's arrival was greeted by over 17,000 Tahitians (half the population at the time), and following that success, he began to train navigators to develop their own wayfinding techniques. Crucially – as Nainoa Thompson, PVS president and an early pupil of Mau, makes clear – this teaching was not about replicating Mau's navigational knowledge, which would have been impossible owing to vastly different cultures of learning, but about inspiring students to develop their own wayfinding skills that were shaped by the ethos

The arrival of *Hōkūlea* from Hawaii, 1976.

of Indigenous Pacific knowledge. One outcome was the Hawaiian Navigation Compass, a circular star map designed to guide navigators using what they can see on the horizon. More important than the invention of instruments, however, was Mau's lesson that memory was the key to navigation. Only through memorizing the stars, ocean swells, wave formations, coral reefs, fish and birds can one start to become a true navigator *with* the Earth.

Today, the pvs launches frequent journeys across the Pacific Ocean as it trains the next generation of seafarers. Since its foundation it has covered 450,000 kilometres (280,000 mi.) of ocean and trained young people from across the Pacific region. By doing so, it is not only keeping a tradition going by reinventing it but ensuring the past is foregrounded in the future. One of the key lessons from the pvs is that traditions in Indigenous navigation can be lost if no effort is made to save them, and this is no easy task in the face of a growing dependence on navigational technology and the move towards universal (read Western) knowledge systems. The task is not to reject these, either – the pvs uses GPS to ensure the safety of crew members, for example – but rather to integrate them into navigational practices that are grounded in the ethos of wayfinding *with* the Earth rather than *on* it.

In both the First Nations and Pacific Islander examples, the maps that are used do not correspond to a common understanding of what a map is, nor to a Western understanding of how spatial knowledge is shared. The maps in question are not even graphic, but these are mapping practices nonetheless. They show how history and culture matter in our reading of spatial knowledge. There are distinctly different ways of reading these maps and of passing on geographical knowledge to others. Whereas some maps can be passed on freely with a shared understanding of how they might be used, Indigenous mapping practices must be learned and earned through lengthy periods of practice and respect for where this knowledge came from. There is no given right to spatial knowledge and how to practise it. This is a challenge to the notion – which continues to be taught throughout the world, but is not shared by all – that one can simply pick up a map and use it to go anywhere. Maps and world views go hand in hand, and it is important to be cautious when making universal claims about how geographical knowledge is produced and shared with maps. As we shall see, context matters.

Mapping contexts

On 1 October 2019 President Donald Trump was facing an impeachment trial. Using his favoured mouthpiece, he tweeted a political map of the United States indicating how each county, in every state, voted in the 2016 election.[15] The map is mostly red, suggesting unequivocal Republican support for him and his administration. Overlaid in bold white type is the tag line 'Try to impeach this.' The message was clear: he had the public's support across the vast majority of the United States, enough to ride out the impeachment trial.

Trump was right to think that he would remain president, and on 5 February 2020 he was acquitted by a party-line vote from the Republican-held senate. He was wrong, however, to assume the map posted on Twitter (now X) showed overwhelming support from voters. All it showed was which way each county voted, not how many people voted for him. Some counties have very few people living in them. For example, LA County in California, represented by a sliver of blue on the West Coast, has a population of almost 10 million, whereas the whole of Kansas State, represented by 105 counties in a sea of red at the centre of the map, has a combined population of just 2.9 million.

What the map does not show is that in 2016 Trump lost the popular vote to Hillary Clinton by nearly 3 million votes. Instead, it shows that Trump is an ultra-savvy propagandist, who knows that maps can tell tall stories and influence what people think. This map, in common with electoral area maps all over the world, misleads us into thinking that blanket coverage on the map must mean popular support. Had Trump tweeted a dull statistical table of the overall voting numbers instead, an altogether different form of representation, his national support would have appeared far less impressive. The table would have read: 62,984,828 votes for Trump and 65,853,514 votes for Clinton.

Irrespective of the facts, Trump's use of the map to distort the reality of his support proved to be an effective way to create a believable illusion about the country. To his supporters, and perhaps to Trump himself, the map showed an accurate representation of a country that would swing behind its embattled president. This map is not simply a two-dimensional representation; it is a political act of propaganda spun from the hands of the infamous late-night Twitter user and intended to shape the political beliefs and actions of the American public. It is this context that gives the map its meaning. Without Trump, his known tendency for having used Twitter as a public mouthpiece, and the rapid spread of information made possible by the platform, this map would remain a graphic representation, but it would lose all its significance.

Context also matters beyond the global scale of national and international politics. Take the redevelopment map used to show how the Elephant and Castle regeneration project in south London will look by the time of its completion in 2025.[16] This interactive digital map is presented online and at public consultations about the project. For the development organization, the Elephant and Castle Partnership, it is a tool to illustrate the final development for investors and future residents. For them, the map shows plainly how this long-run-down area of the city will be improved. In contrast, for the South American communities being displaced by the redevelopment through rising rents and unaffordable business taxes, the same map may become a symbolic reminder that someone is coming for their homes, their businesses and their communities.[17] For these people, the map may be a powerful illustration of the gentrification that is displacing them, and not the alluring image of a prosperous urban future that is surely intended by the cartographers behind it.

This kind of redevelopment mapping is popular among developers and made mandatory by local planning authorities. It is often placed alongside fictional and highly aestheticized computer-generated images that give an artistic impression of how the site will look when it is finished.[18] These images feature the kinds of person and practice to which the developer is marketing the properties. Workspaces, play spaces and leisure spaces are stitched together under blue skies, where everyone and everything works together to create harmonious space. In effect, an urban future is ushered into existence through maps and dazzling digital imagery. At the same time, an existing urban reality is erased, starting with the notable absence of those people who currently call the area home. This reimagination of space is a design tactic used deliberately by developers and local planning authorities to legitimize profit-making landgrabs of valuable inner-city plots at the expense of the communities that are deeply rooted there.[19]

This is nothing new. When Georges-Eugène Haussmann 'rebuilt' Paris in the nineteenth century, installing the grand boulevards that are associated with the city's architecture today, maps were used to simplify and cleanse the destructive process of bulldozing the lives of the poor and the places these people called home. The maps used in this case, in a similar way to those at Elephant and Castle, were not *just* maps; they were maps used in the context of state power, where the cartography on the page was used to bring to life Emperor Napoleon III's dreams of a new Paris free from the perceived blight caused by the proletariat, a city that channelled bourgeois importance and grandeur to the rest of the world.[20]

In this localized context of cities and districts, maps and the imagery that goes with them matter to the way people see themselves in relation to the places in which they live and how planners see their work being put into practice. Without an understanding of this context, the redevelopment map of Elephant and Castle could be seen as just another interactive online map to play around with; its significance would be lost. This and the Trump case highlight how context shapes the way people see and react to maps. There is no universal reaction to a map. As maps circulate in society, so do a wide range of interpretations and impact. Some of these are matter of fact, pragmatic or ambivalent, others can be highly emotive, enthusiastic or engaging. Whatever the response, a map seemingly comes alive during the point of interaction. I call these points *mapping interfaces*, where maps and people face each other, and interactions unfold.[21]

Despite all the research that has gone into maps – and there is a lot of research – this line of enquiry is under-studied, in favour of examining the maps themselves. This is especially true in the discipline and industry of cartography, where the focus remains on how to produce the most accurate and aesthetically pleasing representations. But it is recognized by human geographers and others that there is value to studying how maps are used and understood, and by whom, as well as what they show. Rob Kitchin, Martin Dodge and Chris Perkins note that the

> cultural turn in academic Geography encouraged
> a growing emphasis on the contexts in which maps
> operate, encouraging a shift away from theorizing
> about production and towards philosophies of mapping
> grounded in consumption. Here, the map-reader
> becomes as important as the map maker.[22]

Digital mapping

Key to studying maps in this way today is recognizing how digital technology plays a major role in how they are made, read and shared. After centuries of making and using analogue maps from paper, wood, stone and sand, there has since about 1970 been an unprecedented acceleration in the use of digital maps and digitalized mapping techniques. During this time, Geographic Information Systems (GIS) have had a profound impact on the production of maps, changing the way in which they are made and by whom. Similarly, imaging and sensor technology has transformed how geographic data about the world is collected. This shift has accelerated even further in the age of GPS and the smartphone – which, at the time of writing, is barely twenty years old. During this time we have seen the unstoppable march of egocentric maps, which are designed around the blue dot that represents the user, and location-based services, which include all manner of digital services for use at specific locations.[23]

In the digital age, scholars have had to rethink the map. Navigating with other people is one thing, but navigating with digital maps that can pinpoint location and update us on the most efficient route in real time is another. There is a prevailing assumption in society that digital maps do what analogue maps did, just better and more efficiently. Barely a month goes by without a new mapping app or service entering the market, promising to make our lives better by offering advanced routing algorithms, search functions and intuitive user experiences. Digital maps are sold as essential tools that offer seamless experiences and never allow us to get lost. In reality, research shows that although digital maps are improving constantly in accuracy and design, they do not always live up to those promises.[24] We have all cursed our phones when the map has failed to get us to where we want to go, poured scorn on the mini-cab driver who has got lost after following their phone's directions rather than ours, and lost patience with the pinch-to-zoom function of digital maps. This has not gone unnoticed by researchers who wish to understand how the intricate details of map use shape people's lives.[25]

Research also shows that we are no closer to becoming a society of expert digital-map users, no closer to being able intuitively to understand the world in new ways through the shiny mapping interfaces that are now omnipresent.[26] Much of this research has suggested that people may have become poorer navigators, far worse at locating ourselves and places on the map without the aid of a search bar, and less able to understand our location in relation to others. With all the good that GPS technology and search functionality has brought in being able to locate ourselves on the map, questions are being raised about how egocentric maps, in particular, are affecting perception of the wider world.[27] These are important questions to consider if we want to understand how GPS has affected our ability to navigate, but they are equally important for understanding a world that is becoming more global at the same time as countries and people are becoming more self-centred. Maps about me – produced just for me by corporate actors – are not detached from the libertarian notion, and politics therein, that this is a world built around *me*.[28] We must see how maps are changing as a response to how the world is changing.

Rather than perpetuating the myth that the digital map is simply better, the chapters in this book show that although the technology of maps has changed, and that this has had an effect on how and why they are used, it is a mistake to assume that digital maps have revolutionized understanding of how spatial knowledge is represented or used. There is perpetual hype for digital mapping technology that often does not fulfil its potential, encumbered as it is by bad design, problems with connectivity, inadequate customer service and an ethics that does not extend beyond an ethos of 'West knows best'. I want to resist the technobabble of Silicon Valley, from which most of these iterations come, and instead put the digital map into contention with the messy reality of everyday life.[29]

Digital maps may be everywhere today, but analogue maps have not gone away. On the contrary, sales of paper maps are on the rise. Ordnance Survey, the UK's national mapping agency, sold 1.73 million maps in 2017/18, a rise of 7 per cent on the previous year. In the United States, the publishing industry trend monitor NPD BookScan estimates that sales of paper maps and atlases increased by a compound growth rate of 7–10 per cent per year between 2015 and 2020. Similarly, the pedestrian and transport maps that adorn the pavements of towns and cities have grown in popularity as urban centres create 'pedestrianized zones' in ever greater numbers. Analogue maps remain stubbornly and curiously entrenched in a society convinced that the digital revolution is here.

There are many reasons for this. Analogue maps offer different options for the travelling user. Rail maps attached to station walls and inside train carriages are often quicker and more efficient to use than fiddling with the same map on a screen. There is a reason Harry Beck's famous design for the London Underground map, despite many alterations, has stood the test of time as a large-scale wall map *and* a pocket-sized printout. My own research into map users in London has found that people tend to pick and choose the map that is the most convenient for them at the time, whether it be on a phone or on the wall.[30] The brief glance given to Beck's map on a crowded Tube train, or the detailed analysis made by a tourist, is often much better suited to a large printout than it is to the screen. It also invites a different kind of interaction, one that makes our plans visible to others as we trace a route across town, or audible as we discuss it with our fellow travellers or seek help from a stranger.

Another reason is that analogue maps do not run out of battery or lose signal. They are, in a sense, more reliable even if they could never be as up to date as the digital alternative. Such stability is often why paper maps are chosen over their digital counterpart by adventurers, and why mountain-rescue services advise leisure seekers not to rely on their smartphones. It has also become part of the backlash against digital technology, often playing a role in the rise of 'digital detox' movements that demand we unplug and find non-digital ways to engage with the world. To buy and use a paper map for a weekend's hiking is becoming a statement of intent *not* to engage with the digital technology that is so prevalent elsewhere in our lives.

Defining the map

All Mapped Out explores the idea that maps and what we do with them cannot be universally defined. I want to highlight how ideas about maps frequently clash with the reality of how and why maps are used. By bringing together research from my own fieldwork studying map users and that of others who have studied and written about maps and mapping practices around the world, I want to show how views and uses of maps are shaped by different cultures, communities, contexts and technology. It's high time to rethink the many ways in which maps are used. There remains a prevailing view that maps are neutral and objective, once paper and now digital, accurate and functional, despite the now well-used line that maps are arguments made about the world.[31] Why is this? And how do we move beyond it? I use this book to suggest looking beyond the map, towards the varying contexts in which maps are used and the variety of people who make and use them.

I believe maps are representations tied to spatiality, whether they be geographical, pictorial, political, thematic, diagrammatical, cosmological or spiritual.[32] I also believe they shape and are shaped by all kinds of social, cultural, political, economic, environmental and ultimately spatial practice. I do not think maps must necessarily conform to the conventions that we tend to hold in our mind's eye, nor to the totalizing view that maps must be all about objectivity and power (the common dictum of the field of critical cartography). Instead, I like to think about maps with the late Denis Cosgrove's idea of 'mappings', which he used as a concept for all that represents spatiality in one way or another. He states: 'The measure of *mapping* is not restricted to the mathematical; it may equally be spiritual, political or moral. By the same token, the mapping's record is not confined to the archival; it includes the remembered, the imagined, the contemplated.'[33] Through this framework, we are given much more freedom of thought around what constitutes a map and what can be done with one. It helps us to untether our ideas about maps from the reality of their use and impact.

The common understanding about maps is not wrong. Maps *are* useful navigational aids, powerful objects of geographic representation and political devices – we will get to that – but they are so much more. They are also cultural objects with significant histories, ideas about the world and how it is understood, collectibles for obsessives and artworks for subversives, things to bind people together or tear them apart, and objects in and of themselves, which collect dust, are scribbled on, walked on or used as temporary coasters. There are many ways to read or use a map, and these include but also go far beyond what is shown on the face of a map. How a map is interacted with often depends on the person, where they are, what they are doing and who they are with.

One aim of this book is to present evidence that begins to acknowledge that there is a wider context to consider. Through developing this argument, I build on a growing interest in thinking about maps beyond what they represent. This extends the established fields of critical, social and historical cartography, from which the majority of studies about maps have emerged.[34] Coming at maps from the perspective of what people do with them, and how they affect them, offers another way to understand this ubiquitous cultural object.

I want the reader to keep this simple question in mind over the following pages: what are maps? This question probes at the common underlying assumption about maps: that we know what they are and why we use them. All we can hope for in the pages of the dictionary is a vague description, one that – as is nearly always the case – barely begins to explain the worldly significance of the object in question. The *Oxford English Dictionary* (OED) defines a map as 'a drawing or other representation of the Earth's surface or a part of it made on a flat surface, showing the distribution of physical or geographical features'. I suspect most people would agree with this. *I* agree with it to a certain extent. The definition tells us that maps can be summed up in one sentence; that maps are *simply* a representation of the Earth.

The definition is useful because it works.[35] It helps us to differentiate maps from other forms of geographical representation, such as the written word, landscape painting or photography. Nevertheless, it does very little to help us understand what maps are for, whom they are used by and why. It tells us something about how society sees the map, but almost nothing about how society uses maps. 'Showing the distribution of physical or geographical features' is suggestive of what we might use maps for – navigating between these features, for example – but it is hardly an all-encompassing descriptor. What about the scenes that unfold during navigation, where the map can play a role in the drama of getting from A to B? Or the ways that maps might shape the perception of a relationship to those geographical features? Maps are, after all, evocative objects that can spur our imaginations, both in literature and when used in a soggy field.[36] The world of surfing is a case in point. Swell maps are cartographies that are consulted regularly to determine when and where to catch a decent wave, but they are not used simply to pinpoint the location of swells; they are used to build excitement and anticipation for the adventure ahead.

Dictionary definitions offer a matter-of-fact description of a category of loosely similar objects. The same generalization could be made for 'books', 'paintings' and 'photographs'. However, while no one would say that these are all the same – there are many categorizations for each – many people have come to the conclusion that all maps are more or less the same kind of object, with the same purpose. A universal set of assumptions about maps has emerged in ways that differ from other forms of representation, which is odd considering the popular practice of categorization that is applied elsewhere. Certainly, we categorize maps loosely, based on which place or data they represent, or what activity they are associated with, but we hardly unpack them in popular culture in the same way we do books, paintings and photographs, where tremendous effort goes towards arguing over artistic style, period and value.

It is easy to fall into the trap of assuming that all maps are the same and always have been. Using the OED definition, the rock carvings of Bedolina, the modern plans of London, the route planners of the New York Metro, the sea charts of the Pacific and the satellite images of Google all share the same aim. They are all objects that show physical or geographical features of the Earth's surface. But if we take a closer look, they are neither the same type of representation – there are vast differences in scale, orientation and detail – nor used in the same context. What the OED definition fails to recognize is that maps do different things for different people, and always have done. This is recognized in academic research, but rarely acknowledged in the wider discourse.

We have learned to trust and believe in the objectivity of maps because they show us the places we have been to, the places we want to go to and the places we don't or never can go to. They have helped us to get there, literally and figuratively. They make understanding the world simple: *this* event is happening *there* on the map. I am *here*, because there I am, *on* the map. But do we really know what maps are and what they are for? Do we ever question how they shape everyday life – where we go and the experiences in getting there? Do we really know anything about how other people use the same maps as us, or consider how different social situations, cultures, places, times and digital devices might change how or why we use the map or how the map makes us feel? This book sets out to answer these questions, with the aim of showing that there is *always* more to the map than meets the eye.

My first example, the Bedolina case, showed how assumptions can quickly be challenged when today's understanding of maps is plotted on to the past. Following this with examples from non-Western and Indigenous mapping practices, I suggested that assumptions about how maps look can be questioned by looking beyond the conventions of gridlines, scales and systems of coordinates, and to other cultures of mapping. In the following chapters there are many more examples to show that maps are far more diverse in appearance and use than we might think. By laying out these cases I will show that maps have an important role to play in who we are, where we have come from and where we are going. By examining the complex ways in which they shape our lives – from the politics they create to the emotional responses they evoke – I aim to show that our understanding of maps cannot be boiled down to the cartographic form they take. Instead, I'll argue that maps are what we make of them.

Mapping out the book

All Mapped Out sets out six overlapping themes to chart how maps have become tied to culture and society. I want to do more than identify maps that have changed the world, or lay out the history of maps and society, since this has been done over and over again. Instead, I intend to show that all maps have the potential to change the world and shape society. It's just a matter of where you look and whose world you are interested in. Each chapter is intended to inspire another look at maps, rather than to be used as a definitive guide. I am not the authority on maps or the impact they have in the world. The reader will have many other ideas from their own knowledge and experience about how or why maps are used. The point is to spur on a growing conversation, which is so far being had only in a small corner of map studies, and which reaches beyond a matter-of-fact assumption about maps and how we use them.

In Chapter One we look to perhaps the most common use for maps – navigation – to show not how it might be better understood by looking at how accurate or user-friendly maps are, but how this set of practices is shaped by specific social situations. I will argue that it is wrong to assume that we all navigate in the same way. While some of the quirks of navigation may be amusing or idiosyncratic – such as the character who carries a compass to find north when using his smartphone, or the friend who pretends she is on the run from the police in order to plot the most efficient route – there can be serious implications for believing that assumption to be true. This chapter shines a light on navigation as a social and cultural practice in which who we are, where we are going, who we are with and what form a map takes can affect how navigating with a map unfolds.

Chapter Two focuses on maps as interfaces for understanding movement. Maps are the common interfaces that we use to view and make sense of movement. During the COVID-19 pandemic (2020–) and the conflict in Ukraine (2022–) this came to the fore as maps were used as a primary tool for showing where infections and insurgencies were spreading, and the extent to which they might affect us and others. But there is a much longer history to this type of map, and there are far more ways in which we can understand how maps might shape movement. By telling a story of maps and movement and how we came to trust maps as the primary means of recording and viewing the movement of people and things, this chapter shows how maps have shaped our perception and under-standing of movement at global and local scales, as well as how these interfaces affect our personal lives and embodied experiences.

Chapter Three seeks to go beyond the now well-known argu-ments about the power of maps, which is to say that they are powerful social constructions that have shaped society throughout history. Taking this as a starting point, this chapter repositions the map–power relationship as inherently unstable, and uses an analysis of Google Maps and maps used during the European migrant crisis to argue that a true understanding of the power of maps requires us to examine the situated ways in which they are produced, used and circulated.

Chapter Four examines the relationship between maps and culture. For as long as people have made maps, cultures have been represented on them. But, since the move to make maps instruments of science in the eighteenth century, they have become regarded as objective, and are rarely discussed as cultural artefacts outside map studies. This chapter starts from the premise that all maps represent the cultures in which they were made, and looks closely at how different cultures are represented on the map. I then turn from cultures on the map to focus on cultures *of* the map, to the people, groups and communities that have formed a social bond based on what they do with maps. The intention is to question what we can know about culture by looking at maps, and to understand what we can know about maps by looking at the cultures in which they are used.

Chapter Five is about maps and money. It tells a story of how maps have long been tied to economic activity, and how digital maps and location-based data have reconfigured this relationship significantly over a very short period. It first asks questions about what cartography produced by colonial and capitalist economies has done to our relationship with the value of land, and how these might differ from other ways of seeing the world. Then it focuses on the rise in the economic value of location-based data that is visualized on maps. Through a discussion of data and maps in the digital age, I engage with debates about freedom, privacy and surveillance at a time when our location is closely monitored by the devices that rarely leave our side. Together these stories point to the importance of following the money if we want to understand what maps are used for.

In the final chapter I look to the future of maps without buying into the hype around mapping technology and what it will do for us. Using the emergence of machine-learning mapping technology and geo-visualization, which are applied in industries ranging from self-driving cars to archaeology, conservation and urban planning, this chapter asks what becomes of maps when they are not necessarily designed to be read or even seen by human eyes, and what new questions this asks of maps and how they shape the world. It will also explore the tension that emerges between the hype of advanced mapping technology, the messy reality of complex global issues and the nuanced ways that we use maps in our daily lives, arguing that so-called revolutionary shifts in mapping cannot be marked so easily by a narrative of linear transition, from one map to the next.

By refocusing attention on the different spaces and places in which maps are made, and whom they are used by, I aim to offer a different perspective on the work maps do in the world. When studied with personal histories, social situations, political differences and different environments in view, maps and mapping practices take on new meaning: a meaning that goes far beyond that which we tend to hold of the map as the standard for spatial representation.

1

Navigation Beyond the Map

Neuroscience would have us believe that navigation is a purely scientific process that happens when the hippocampus interacts with the pre-frontal cortex in the brain. The science tells us that we build up a mental map of the places we have been to using our so-called place cells and grid cells, first identified in studies of rats. Through repeated visits we develop a deeper understanding of places and become better at navigating them. This is why we have a detailed understanding of the areas we grow up in, why we know all the back alleys and shortcuts there, and why so many of us struggle to locate ourselves in places unknown.[1]

London's famous black-taxi-cab drivers, who have long claimed that they know the streets of Britain's capital better than anyone else, have become the go-to guinea pigs for the science of navigation. Experiments in which drivers are tested on their spatial knowledge of the city and then put in a functional magnetic resonance imaging (fMRI) scanner prove that cabbies develop a larger hippocampus than the average person.[2] We are told that the hippocampus has grown to accommodate the skill and memory needed to drive around London's famously muddled streets without the need for a satnav. This may come as no surprise. Through years of studying the map of London, driving endlessly around the city and completing the notoriously difficult 'Knowledge of London' test, in which drivers must speak out the most efficient route between two points given by an examiner, they have earned the right to claim they know the city well.[3] The neuroscience simply supports what they and many a Londoner have known for some time.

In my study of taxi drivers navigating London, I spent some time with Dennis, a black-cab driver for more than thirty years. He certainly knew his way around town, and I have little doubt that an fMRI scan of his brain would reveal similar results to those mentioned above, but I wanted to know more about his day-to-day navigational processes. Did he ever use a map? What kind of map did he use? And how did his mapping differ from the mapping of other people? What I found was surprising because it challenged the myths about black-cab drivers being *the* experts of route-planning.

Before smartphones, many a Londoner and visitor alike bought a copy of the *London A–Z*, an iconic pocket atlas, to get around. Dennis still used this map. He had a tattered copy stuffed into the pocket of the driver's-side door, but these days he preferred to use the electronic version because it updated automatically, saving him from going out and buying yet another paper edition. Having learned 'the Knowledge' using the paper A–Z, he was no stranger to using it. He explained that most of the time he didn't need to, but that it remained an essential tool for the job. In between fares he would scroll through it after coming across changing road layouts, unforeseen roadworks and popular new tourist attractions. Many years of driving around the city had compounded his core knowledge of the road network, especially in the city centre, where he spent most of his time taking fares, but he still had to keep up to date with the city's perpetual changes.

For large parts of the day, knowing how to get somewhere was not a priority, or something that Dennis found especially difficult. He did this on autopilot. On one of our journeys together, we reached a road blocked by stationary traffic ahead. He made an almost instant calculation and we headed down a side street, one of those streets that you would never think of taking but are always curious about. He knew straight away where it would lead us, and this quick decision resulted in us taking an entirely different route from the one he had planned, as if it were instantly plotted by an algorithm, taking into account how long the extra route would take, estimating traffic and cost, all in a single moment. At the very same time, the other road users retained his focus, as he made sure not to bump into another vehicle, take out a cyclist or run someone over. He barely flinched when a bus tried to swoop in front of him, or when a group of teenagers ran across the road without looking. This of course is simply an example of a normal day in the life of a taxi driver, but when you stop and think about it, it demonstrates amazing cognitive skill.

Nonetheless, there were occasions when Dennis did have to pause and think carefully about where he was being asked to go. It was at these times that he recalled his training with the map. In his mind's eye, he had an image of the Greater London area that looked like the A–Z map. This image, he said, was what he had built up over years of working with the A–Z atlas, the wall map and the dog-eared printouts he made for his moped, which he used as a training vehicle while learning the city (still a common practice among learner cab drivers). When he was asked to go to an area he had not been to in a while, he used this image and the A–Z app to jog his memory of possible routes, and he did this by the way it created a patchwork of distinguishable shapes of London's road network. The GPS coordinates system used in navigational apps was of no use to Dennis. He divided London into large and small shapes that represented different road networks – major and minor roads, key junctions, bottlenecks – and significant places. He used these as key reference points for navigation by linking them together as he drove, ticking them off as he passed through them.

One of the experiments Dennis and I did together was for me to call out random destinations and for him to explain the process as he drove us there. On a trip starting at Angel Tube station in the northeast of the city and ending in South Kensington to the west, a journey of about 8 kilometres (5 mi.), he recalled how he first thought of the destination as a key tourist site and then as a road layout based on a rectangle containing universities, museums and cultural attractions. His second thought was Hyde Park, because it was the largest shape on that route – represented on the map as a large green rectangle – and essential to pass when getting to the area from the east. His third was a semicircle that encased Holborn, Aldwych and Covent Garden. Between Angel and South Kensington, he used these shapes and more to make a personalized route for our journey, one that could be tweaked easily depending on who he was driving or what he was presented with on the way. If he was carrying a tourist looking to see landmarks on the route, he would factor that in. For businesspeople looking for the fastest route, haste would be taken into account.

These shapes helped Dennis to simplify the sprawl of London's road network in order to piece it back together in a way that worked for him. He was not sure if other drivers made routes this way – perhaps it was a quirk that had stuck from his training – but clearly it worked. However, it was interesting to discover that it worked only in London. Take him anywhere else and he would be lost in no time. He would be reliant on asking people for directions, using his smartphone or, most commonly, his wife, who is much better at navigating in new places. Chuckling behind the wheel, completely at ease in London's busy West End, he said that if he were driving in a new place, he would be the one pulling over in a panic every five minutes to check if he was going the right way.

Neuroscience is useful in explaining what is happening in the brain when we navigate with maps in specific places, but as this insight into the daily activities of a taxi driver illustrates, the science tells us very little about how drivers read maps and how the experience of navigating unfolds. We are all similar to Dennis in the sense that we have our own quirks of map-reading that help us to get from place to place, and these cannot be fully recognized when brain activity is monitored in a lab. What the science has yet to do is study navigation as a lived experience. It has neglected to examine how navigation happens in different contexts and cultures, and how different maps and whom we use them with might affect how we get from A to B.

Popular discourse suggests that we are losing the ability to navigate because we blindly follow the GPS. There is some truth to this, and it has been proved in a lab setting alongside anecdotal reporting, but all too often this simplifies the reality of driving with maps. The fact is that we do not know enough about driving habits and behaviour to make substantial claims about the damage GPS might be doing to our cognitive navigational abilities. When we are out on the road, there are all manner of factors that shape our use of GPS devices. When we drive, what we drive, where we drive, how we drive, whom we speak to and what technology we use can all shape our experiences of using GPS. Then there are all the other social and situational factors relating to who we are, how we are feeling and what we are thinking about that can affect these experiences. The science may show that GPS is affecting our sense of direction or our ability to know exactly where we are, but in reality many people who use GPS are not using it to know that. Instead, they are using it as a countdown timer for how long it takes to get there, or, in the case of ride-hailing drivers, to show passengers that they are following the correct route. They may have travelled the route hundreds of times, and know the turn-offs, traffic-light systems and pinch points well, yet they still use GPS devices. Why? Because there are social, cultural and economic factors that are not easily explained by the science of navigation.

Research into navigation is only beginning to investigate how these details might build an understanding of navigational practices in the digital age. This is where the social sciences and humanities have started to make a contribution. Research in these fields focuses on the lived experience of navigating with a variety of different maps. It is also where different ideas about what navigation is, or could be, come into focus. Navigation is a process that stretches well beyond an exercise in efficient orienteering; it is deeply embedded into our ways of life.[4]

Human geographers have found that how you travel, where you are and at what time of day you are there influence how a map is used in transit. For example, taking a bus in an unfamiliar area late at night and using your smartphone to check your location, so that you know when to get off, is a very different mapping experience from using a satnav device while driving to a friend's house in the comfort of your own car. My own research into map users has found that there can be a real sense of jeopardy and danger in the former, as one looks at the map, out of the window and back again. This sense is compounded for women travelling alone, is less of a concern for men, and eases for all when travelling with a trusted person or group. In the latter case, I found the opposite to be true. All the drivers I interviewed expressed feelings of comfort and boredom as the map counted down the miles and warned of upcoming traffic. In both situations participants used Google Maps, the most popular navigation app, but the context shaped the experience of using that map in quite different ways.

Instead of laying out how neuroscience has come to understand navigational processes, or how maps and mapping technology have shaped navigation throughout history (which would take an entire series of books), this chapter focuses on two common navigational practices, driving and walking, to tell stories about the social and cultural side of navigating with maps. We will see how these activities, both of which will be well known to the reader, produce different navigational experiences that challenge the notion that we all use road and street maps in the same way. By showing the wider contexts of how navigation happens and the impact it has, these stories will question the idea that navigation is a practice that unfolds only between people – or a dedicated person – and maps, and the popular idea that GPS and satnav technologies are erasing our ability to know where we are and where we are going.

Driving with the map

Patient calls to 'Be quiet in the back seat,' turning to screams of 'Shut up!', are firmly embedded in my memories of childhood holidays, while my mother concentrated on the map and my father drove us around in circles on our annual trips to northwestern France. Tension would flare when turns were missed, or when navigational instructions were not conveyed quickly enough. Frequent stops were made to clear heads, storm off, refocus and get our bearings before we ploughed on. Tantrums calmed momentarily. Sitting quietly at the centre of these situations was the Michelin road map, seemingly innocent and resolutely practical in all its spiral-bound glory. It was simply the tool my mother used to get us to where we wanted to go. Or so we believed.

In fact, this map was not as inanimate as we might have thought; we should have taken it more seriously as an object that shaped our holidays. Today, I would argue that this map was thrown into our family context, animating itself through us in its shaping of our family life played out on foreign roads. That same map, used by another family on those same roads, would have different experiences to report, perhaps less fraught with tension owing to a more harmonious navigational partnership between the parents.

Driving and navigation are practices that go hand in hand. This is nothing new – drivers have used maps and navigational devices since the widespread adoption of the car[5] – but digital technology has in recent decades changed what it means to navigate.[6] For many of us, it has also changed whom we turn to for navigational instructions. No longer are we necessarily looking for direction from a friend or partner in the passenger seat, sitting with a road atlas in their lap, or stopping at the service station to check our route. Now many of us simply get in the car and type our destination into our onboard GPS device or smartphone. But this is not the whole story. We have not all become robotic followers of the maps that adorn our dashboards, despite reports of drivers following their satnav off the road and, sometimes, off a cliff. Nor have we all succumbed to digital navigation devices; paper maps, road signs, landmarks and directions all still have a role to play.

Rally-car racing offers a window into a world where effective navigation still relies on paper. A rally driver relies on their navigator to win races and avoid high-speed crashes. Over the course of a racing event, co-drivers build up an intricate knowledge of the course that is conveyed to the driver through clear verbal instructions based on a series of 'pacenotes', which we can consider a type of mapping. These notes come in the form of a spiral-bound notebook, handwritten or printed (and easy to flip through), or are displayed on a tablet using a dedicated app.[7] How each page looks depends on the partnership and team in question, but common features include details of corners, directional arrows, small maps and short pieces of text indicating distance, gradient, time, speed, obstacles and gear changes. Similarly, the way each page is read differs significantly according to a number of factors, including the driving partnership, the language and how long the pair have been working together. In this corner of motorsport, shorthand vocabularies of navigational instructions have emerged, which are in stark contrast to those spoken between the average driver and passenger, or to the average driver by a GPS navigator.[8]

A rally driver has a nuanced understanding of how to respond to their partner's notes, which can take many years of working together. In the most successful partnerships, there is a union between drivers, car, course and pacenotes.[9] There must also be trust across the union for the sake of keeping each other safe and the car on the road, and ultimately winning the race.[10] The key to a successful rally is the reconnaissance – or recce – of the course that is done before the stage race begins. Four weeks before the start of each racing fixture, the organizers give the team a road book containing general information about the course, including course maps, a key for track signs and a basic set of routing instructions. During the week before the race trials, each team is permitted two runs of the course in normal driving conditions, to develop their pacenotes.[11] On the first run, drivers read out navigational notes to the co-driver and a rough draft is made. On the second run, the co-driver reads the notes back to the driver, who calls out further notes while driving, helping to produce the second, more fine-tuned draft. Throughout the race weekend, as conditions change or the car is tweaked, further iterations are produced. Moreover, the whole process must be done anew each season, because it is specific to the condition of the course and must take into account changes in the layout; these maps cannot be shelved for a year or so, as others can. This is also why a pencil, eraser and sharpener remain a key part of the toolkit for many co-drivers, in addition to the more modern inclusion of dashcams, which record footage that can be used to make changes to the notes later.

Driving the course in race conditions is a high-pressure task. The pacenotes must be read aloud and clearly, with a specific cadence. Depending on where they are on the course, the driver needs to know in time what the next instruction is. Being too early or late with an instruction, mumbling a cue or giving the wrong command can have catastrophic consequences. It requires tremendous skill and concentration from the co-driver and great trust from the driver.[12] Pavel Dresler, a co-driver on Team Škoda Motorsport, notes:

> Simply put, the driver has to drive exactly as I tell him
> … A good co-driver's job involves much more than just
> reading while the car is in motion and not getting travel
> sick in the process. He has to have a great sense of rhythm,

A set of pacenotes. Each line of the shorthand on display here refers to the grade of a straight/corner (slow to fast), the duration of a straight/corner (short to long) and a description of the road surface and surroundings (for example, bumps, dips or jumps in the road, an uneven road surface or an obscured view). Co-drivers must learn this shorthand and how to deliver it effectively to their driver on the road.

be in perfect shape, know how to drive and understand the technology under the bonnet. And, when it boils down to it, he's a writer, too.[13]

Rally driving is not especially well suited to digital navigational equipment. Although today's GPS technology could easily represent the route and course to the driver, it would be well off the mark in first recognizing then conveying the contextual information to the driver at speed. Such information as the way corners become more or less acute when driving around them, or how the road surface differs upon entry or exit to a corner, is not to be found in the instructions given by Google Maps and the like.

Nevertheless, attempts have been made to digitalize and automate this process so that at the start of a race meeting each driving team receives the same set of pacenotes, which they can use as a base map. One software program, the Jemba Inertia Notes System, uses sensors to track and record acceleration, braking and topographical information to produce a set of pacenotes, but nothing has yet replaced the co-driver.[14] Even with such systems in place, additional work is needed by the co-driver to tweak and hone the notes before race day. This could be because of limitations in the technology, which is not able to cope with the complexity of a rally course and varying driving styles.[15] It could also be caused by the fact that this does not represent a commercial opportunity for the makers of such technology – rally racing is hardly the market that Formula One is – or simply because the driver/co-driver relationship is ingrained in the history and practice of the sport, and still offers a competitive advantage when it works well.

Most people are now used to working with digital navigators while driving, but in this context, navigation must be a human-led interaction if it is to be successful. The social side of navigation matters, too, because rally-driving partnerships are about shared experiences and working together *with* the map. This is not unique to rally driving, nor indeed to driving in particular. It matters deeply with whom you are driving, walking up a hillside or wandering through foreign streets, and how well you know them. It pays to think about your navigating partner in all kinds of ways, from how well they can give instructions to how much you trust them to get you both to where you want to go, and how easily you can endure if you get lost.

Walking with the map

Walking provides a different lens through which to view the nuances of navigating with the map. Together, walking and navigating have a long and complex history that spans millennia and far predates the present-day obsession with driving. Although there are some written historical accounts of walking and of specific forms of walking and navigation, such as the history of orienteering, there is no one historical trajectory of this marriage that can be laid out here, because there is no linear path to how the two practices became conjoined.[16] This is because both 'walking' and 'navigating' are linguistic vessels for all manner of wayfinding activity.[17] There is no useful definition that can encapsulate that.

With this in mind, maps *have* played a fundamental role in shaping different forms of walking and navigating, from where we walk to how we walk, why and with whom. As with driving, there are numerous examples of how walking might bring about different uses of the map in different situations. Walking to the shops, rushing to the station, moseying around a museum or gallery, ambling through a new place in order to get your bearings, hiking with friends in the mountains, strolling to take in the sights, power walking to stay fit and dawdling to waste time are but a few examples that might be shaped by the use of a map.

Here I focus on walking and navigating in cities, and walking and navigating with others, two broad scenarios that many readers will be familiar with.[18] By doing so, I want to highlight how we get around with a map, and how these actions are influenced by the city's physical structure, social life and layered history.

The pocket atlas that fits in a bag, the wall maps that adorn information kiosks, the transport maps on display at metro stations, bus stops and bike-hire docks, the street maps printed on the backs of leaflets and those featuring in guidebooks are just some of the maps that people use to get around cities. Some of these have been used for centuries, and live on today, although since about 2010 we have become accustomed to using smartphones to help us get around.[19]

The maps on our phones have quickly become the first thing many of us think of when we need to get somewhere, but that is not to say that these older maps have fallen by the wayside. If anything, they have increased in use and prominence – we have never had so many maps to choose from. Nevertheless, the digital, personalized map has set a new standard in helping us to find our way because of one defining feature: their ability to locate themselves and, by proxy, us, *on* the map and *in* the city, in real time, using GPS technology. This well-documented shift, which makes *me* the centre of the map, marks a moment in the history of navigating with a map because it changes how we orientate ourselves in the city and how the city orientates itself around us.[20] Despite these changes, navigating with the map is still largely regarded as a purely pragmatic activity that unfolds between user and map, only now done more efficiently through advances in handheld technology. What we don't often consider is the wider social, historical and environmental context in which the smartphone map is used for navigation in the city.

In my study of people walking and navigating with maps in London, this became ever clearer as my participants all declared complete reliance on their phone's maps and yet were frequently observed taking navigational cues from all over the streetscape.[21] One of the key findings of any observational research is that there is a marked difference between what people tell you they do and what they actually do. In this case, signs, street furniture, road layout, landmarks both iconic and unremarkable, advertising boards, the number and destination on the front of buses, heavy and light traffic, the weather, other pedestrians, other maps and other apps were all seen to influence how my participants got from A to B while using the map on screen. These features of the urban environment and twenty-first-century life were cues for my participants to look at the map and check they were still following the planned route. Participants both stopped abruptly in their tracks to do this and did it while walking, trying to follow the route and keep an eye on where they were going. The act of checking the map and especially the whereabouts of the blue dot representing themselves was interwoven with their observations of the urban environment as they moved through it.

How participants navigated which streets to take was also somewhat random. Some made ad hoc decisions from moment to moment. Routes that had been plotted carefully in advance were quickly deviated from as people became interested in their surroundings and what those could offer. Going out of their way to get coffee or a snack, run an errand, take a photo, answer a call, write a message or escape pollution, or simply 'because it's nicer', was common. The result was a constant negotiation between what was presented on the screen and what they wanted to do in the world. Some felt the pull of the map and soon found their way back to their prescribed route, while others exerted more agency, knowing that the map would eventually find a way to re-route itself around them.

These findings reflect a common theme in studies of navigation and wayfinding, where it is observed that people *know as they go* rather than follow a predetermined route to get there.[22] We often have a route in mind, based on our knowledge of a place or on what has been calculated by routing algorithms, but ultimately our movements towards our destination are influenced by what we encounter along the way, be it physical, such as a closed road, a steep hill or a sudden downpour, or social, political and embodied, such as whom we come across, which areas of town we find ourselves in and what notifications appear on our screens.

This view of navigation as a process that is always in flux, rather than a simple map–user interaction, is reflected in other studies that have examined the intricate details of how people interact with the map on their smartphone while walking around the city. Eric Laurier, Barry Brown and Moira McGregor's study of tourists in Stockholm and London in 2016 used portable video recorders and screen-recording apps to observe navigating with the map from the perspective of those doing it.[23] These recordings revealed that the navigational practices of tourists walking in groups were characterized by all kinds of complexity. Tourists were observed talking about the route as they walked, wandering slowly, making abrupt stops, checking their route and reorientating their smartphone map and the 'you are here' dot, as well as taking in and photographing landmarks and scenery. Then there was the rushing, backtracking, changing direction, taking turns to lead and follow, and holding each other's directions to account, both clearly through language and subtly through speech and gesture. The researchers' detailed account of what we do when we walk and navigate in the city as tourists highlights just how complex this process is. The study also goes a long way to challenge the ideas that we *simply* navigate with a map and that digital maps improve the efficiency of wayfinding or nullify the skill it takes to navigate. Instead, it asks us to (re)consider walking and navigating as a social, skilful and experimental everyday activity in which the map plays but one role of many.

Recognizing the importance of the wider urban environment in walking and navigating with a map opens a door to how we might consider the influence of a city's layered history on wayfinding. Clancy Wilmott's study of navigation and digital mapping in Hong Kong in 2020 highlights how using Google Maps in a city that is still laden with colonial presence creates difficulties of all kinds for the local navigator.[24] Through observing people carrying out everyday navigation, such as finding shops, restaurants, markets and city districts, Wilmott describes how the navigator must negotiate a city and an interface cluttered with toponyms that represent the city's past and present landowners. Cantonese, Mandarin and English names for the same places appear all over the street and the screen. On the map, this implies that languages are catered for equally, but in reality, English is the dominant language.

At the time of writing, looking at the map of Hong Kong, we can see that English is more prominent than any other language in the naming of districts, roads, parks, stations, businesses and institutions. This reflects both the continued presence of the colonizer in the city and the West-centric design of the map itself. As Wilmott astutely notes, the map does not represent the messy reality of life on the ground, where places are given different names depending on where they are or who might be visiting them. Instead, it favours the Western view of the city, whereby places must be represented by individual locators on the map and are most likely to foreground consumer practices familiar to the Western (and often touristic) eye, rather than representing places as locations with many meanings, uses, cultures and histories. The result for navigators wanting to search for places outside this gaze, such as for restaurants and markets that are frequented by residents, is that they are difficult or impossible to find on the map, obscured under layers of zoom or recognized only by their English names, rather than by their local, colloquial names.

The phenomenon whereby maps represent selectively what is on the ground has been recognized for some time. Maps and mapmakers have never been in a position to provide a true account of a place; by design, they must choose what to include and exclude. In a world of digital mapping technology, this is encapsulated in the concept of 'DigiPlace', a term coined by Matthew Zook and Mark Graham.[25] DigiPlaces are digital representations of places as seen and experienced through mapping interfaces, but they are far from neutral. The mapping platforms that dominate the market today are in a powerful position to shape how we see and interact with places as they appear on the screen. Type in a destination and you are presented with a view constructed from the interests of the company behind that map. Today that company is most likely to be Google, which returns place searches with maps built around the interests of a Western advertising and marketing giant; multinational companies, chain restaurants, global institutions and tourist destinations are displayed prominently in English.

Independents, lesser-known neighbourhood institutions, attractions and community services are obscured under a few zoom layers, or require more specific search terms. This foregrounds a geography based on Western consumer culture, rather than a geography that reflects the complexity of what is really going on, or the sheer range of activities that people might need to use a map for in their daily lives.

When it comes to everyday navigation, some searches will relate to consumer culture, but many others will not. As Wilmott's study of navigating Hong Kong shows, the dominant street map of the age is not necessarily the most helpful in wayfinding. Going a step further, Wilmott shows how digital maps, especially those with origins in the West, are tied up with the common colonial practice of (re)naming and (re)producing the city to reflect the gaze of the colonizer.[26] She demonstrates how everyday navigation through the city has become bound up not only in the layers of colonial history, but in how the map itself continues to reproduce a colonial view of the city. This leads us to contemplate the deeper role maps play in everyday forms of walking and navigating in the city.

The map is certainly a guiding force for navigators, but, as we have seen, it can be properly understood only by looking at the interplay between map, user and a wide range of contextual factors. This has not gone unnoticed by cartographers and app developers, and many are now designing their services with these factors in mind. The Weather Hi-Def Radar, for example, offers users the chance to plot routes based on the most up-to-date weather conditions. Others, such as the Clean Air Route Finder and AirVisual, help users to avoid the most polluted streets, and a recently introduced feature of Google Maps provides pandemic-ready socially distanced routes that help the user to avoid crowds.

On the surface, these developments are helpful for the navigator because they acknowledge the many ways in which we might approach getting from A to B. But they do have a broader social impact. Maps that help us to avoid pollution, traffic and people might also steer us towards *and* away from routes characterized by different social geographies. Scholars of urban health and pollution have shown that quieter, less polluted parts of town tend to be populated by wealthier residents.[27] Although not the intention of the design, the result is that routes algorithmically calculated as 'good', 'quiet' and 'healthy' become associated with wealthy neighbourhoods – in London, predominantly white – and 'bad', 'polluted' and 'unhealthy' routes become associated with poor areas.

We do not yet know what the true impact of this might be, partly because we do not know how popular these apps are, or how the data that they invariably collect is used and shared. However, in today's economy, which is increasingly driven by data extraction and forms of algorithmic governance, we should not ignore the very real possibility that our use of these apps could inform future decisions about managing different areas of the city.[28] If the data shows that these apps are a popular means of navigation, and if this data is shared with, or sold to, those in charge of managing traffic and pollution, preserving already 'good', 'quiet', 'healthy' and ultimately wealthy areas might come at the expense of letting 'bad', 'polluted', 'unhealthy' and generally poor areas continue to take the brunt of traffic and congestion. Communities living in these areas would continue to suffer the impact of poor air quality and consequently poorer reputations.

This becomes more complicated when we consider so-called safety apps, navigation apps built around designing safe routes for vulnerable groups. On the one hand, such apps can help users to feel safer, which is particularly appealing for people wanting to get to particular places, for particular reasons, at specific times of day.[29] WalkSafe, one such app aimed at women in the United Kingdom, became popular in 2021 after the death of Sarah Everard, who was kidnapped and murdered while she walked home alone in south

London on the evening of 3 March. Its advice for users, to 'Stay alert and stay safer,' is built around a map and GPS technology, and alerts users to recent street crimes as reported to the national crime database.[30] When a user gets close to the site of a reported crime, they are alerted by a notification on their phone. Other features include alerting friends and family at the touch of a button if you feel unsafe, and an option to alert contacts automatically if you fail to arrive at your destination on time. Together, these features are said to create the conditions for women to feel safe when walking to their destination.

But such apps may oversimplify women's experiences of navigation and reproduce the stigma attached to particular neighbourhoods. Kaushiki Das's study of safety apps aimed at women in India found that they universalize the category of 'women' and therefore neglect the significant social and cultural differences that characterize individuals and their reasons for walking in the city.[31] Important differences in class, caste, education, age, region and religion were not recognized in the design of these apps, which included such features as crowdsourced incident reports, photographs and comments about a place. The designs largely reflected the city and safety as viewed by upper-caste, middle-class Hindu women. Calculating routes based on indicators of safety for these women meant that the perception of other women was not always catered to, thus simplifying the way the city is experienced. One woman's 'safe space' can be very different from another's, based on all kinds of contextual factors. This leads Das to pose the question:

> Whose safety concerns are ultimately articulated and
> legitimized and why? Perceptions of safety are contextual,
> and certifications of safety cannot be based on instant
> judgments about a place. The subjective evaluations in
> safety apps tend to present a lopsided notion of safety

which may tilt the scales against localities frequented
by people hailing from minority communities.[32]

As well as failing to take into account contextual differences of
safety, these apps may also curtail individual freedoms. Instructing
the user to avoid an area simply because it seems to be unsafe might
constitute an attack on their agency. It might encourage the user
to stick to well-trodden paths and avoid the pleasures of loitering;
it might diminish the serendipities of place or the chance of adven-
tures in places. Ultimately, Das highlights the irony of safety apps
by showing how they produce feelings of further restriction and
subjugation in women, rather than the empowerment they would
like their users to feel.[33] These apps are not designed to reinforce
patriarchal experiences of the city – where men can go as they please,
while women must watch where they venture – but, calculating
routes based on simplifying the experiences of a group as large as
'women', they could be accused of doing just that. They bolster the
adverse idea that women are responsible for their own safety (rather
than the spaces being made safer *for* them), and that women therefore invite risk upon themselves, venturing out at their own peril,
when stepping past the bounds of the indicated 'safe' routes.

When we look more closely at the databases on which safety
apps are premised, we can see historical patterns of discrimination
repeating themselves. Long-standing sociological studies of crime-
reporting and media coverage reveal that low-income and minority-
ethnic areas of the city appear to be hotspots of crime because that
is where the focus of police attention is and has been historically.[34]
Labelling, targeting and representing neighbourhoods as areas
of crime tends to increase and justify police presence, which in
turn increases the likelihood of their spotting a crime or being
available so that the public can report a crime. This well-known self-
fulfilling prophecy means that crime statistics are skewed towards
places that have always been assumed to be high-crime areas. This

might seem a long way from navigating a safe route home, but the two are closely intertwined. Apps that alert users to unsafe routes or crime hotspots based on this data may inadvertently reproduce discrimination towards historically marginalized groups in society, simply by keeping people out and reaffirming that these are dangerous areas to walk through.

In some cases, this discrimination-by-design has been baked in from the beginning. For example, Good Part of Town (formerly Ghetto Tracker) and RedZone, both navigational apps from the United States operating in the 2010s, have been widely criticized for their racist and classist design, which mirrors and continues the historical practice of using maps to segregate and contain marginalized groups.[35]

Despite the prevalence of navigational safety apps, there is very little research to suggest that they result in safer journeys.[36] This begs the question, why are they still being made? Undoubtedly, wanting safer routes is nothing new. Street crime is a persistent problem, and people have the right to avoid it in any way they can. Unlike maps of the past, safety apps promise reassurance; they appeal to our need to feel safe and offer us convincing real-time data analytics that helps to create that feeling. I am sure that, in many cases, they do work to make the user *feel* safer as they walk. Ultimately, however, we should ask ourselves if these apps create a false sense of security while perpetuating a discriminatory view of certain areas of the city and their inhabitants based on problematic data.

*

Navigation is at the core of what we think maps are for. We assume we all navigate in the same way using the same kinds of map in the same circumstances. This is no surprise, considering the way navigational skills are taught and sold. Orienteering classes at school, map-reading lessons from our parents, YouTube tutorials and glossy marketing videos paint a universal picture of how it is done. This chapter has looked at different forms of navigation to show that this assumption is misguided because it does not reflect the messy reality of our navigational experiences, nor the diversity of the people who are involved in navigating. The 'navigator' in our learning is monocultural, one-dimensional and interested only in the most efficient way forwards. Yet, as we have seen, navigating with a map is a far more complex and social activity, one that is learned and practised in different ways, all of which have an impact on ourselves and how we perceive and move through the world, and on wider society. The next time we pull out our smartphones and type in our destination, we would do well to remember that there is a lot more going on than meets the eye.

For more than three decades, critical scholars of maps have worked to debunk and detail the power of maps and how they shape society, but it is only recently that research has begun to study the ways this power flows through and shapes everyday activities. Even one of the most common daily acts – finding our way – turns out to be a Petri dish for examining the power of maps in practice. We will now move on to other ways in which maps have shaped us and society. Some of these will take a wider view of maps and society, for instance by looking at how maps shape culture, the economy and politics, but the reader should keep in mind the key idea laid out in this chapter: that maps are part and parcel of everyday life, which is where these wider elements are experienced.

2

Interfaces of Movement

In 1877 the German geographer Ferdinand von Richthofen coined the term *Die Seidenstrassen* (the Silk Roads) to describe the ancient trade routes that ran from China to Europe.[1] For centuries these routes have been used to transport goods, people, ideas and information across the largest land-mass on the planet. On his map of Central Asia, which shows present-day Tibet and the plains north of the Himalayas, Richthofen plotted the most common routes going back and forth across the continent. He used blue lines to show the routes of trade coming from China, and red to show the other Eurasian trade routes in operation between 128 BC and AD 150.

In 2013 the Belt and Road Initiative, a multitrillion-dollar project to improve the road, rail and shipping infrastructure between China and Europe, got underway. Billed as the 'new Silk Road', this massive undertaking, spearheaded by China, plans to resurrect the ancient route and create new ones, to cement a modern pathway for goods and people. In its simplest terms, it is both a sign of the ever-globalizing world and a power grab by the world's second-largest economy. A map produced by the Mercator Institute for China Studies shows the current state of the project. The reader will notice how this collection of trade routes is far broader in scope, laid out across land and sea, and passes through many more countries than its famous predecessor. It is an exercise in soft power, whereby the Chinese Communist Party plans to shape and control vast swathes of infrastructure, transport and commerce across the planet. Made centuries apart, these maps share a common theme besides the area and activity they claim to represent. Both shape the perception of the world's superpowers and how they dictate the movement of goods across the globe.

While the maps of old could vaguely show the pathways of trade and indicate which places were the sources of what resources, today's maps can show the precise location of individual items, their current status and, much like a home delivery, their estimated time of arrival.[2] As a result, great economic value is now placed on knowing exactly where things are. This has created a billion-dollar segment of the logistics market, its primary role to keep track of items in real time as they are moved from place to place.[3] This relatively new phenomenon has quickly become normalized, and both businesses and customers now expect to be able to keep track of the things they buy and sell in real time. A wide range of technology – algorithms, barcodes, scanners, pens, paper, transportation

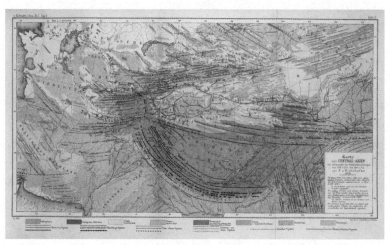

Ferdinand von Richthofen's 1877 map of Central Asia, with different lines showing the routes of trade coming from China and the other Eurasian trade routes in operation between 128 BC and AD 150.

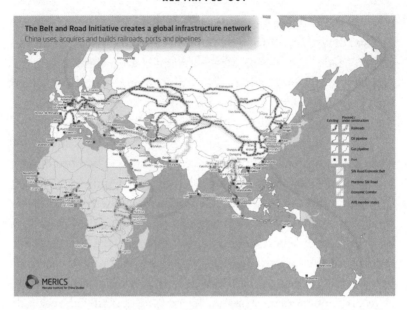

Modern map of the Belt and Road Initiative that will create a new Silk Road running from China to Europe.

and others – is working constantly behind the scenes to tell us when our purchases will arrive or what is holding them up, but maps are especially significant for the layperson because they visualize and make sense of these movements in a way that is familiar.

As well as tracking our stuff, the logistics industry has become interested in tracking and mapping those responsible for ensuring the safe delivery of items. GPS technology is now fitted to the ships, planes, lorries, vans, motorbikes, bicycles and drones responsible for transporting goods. These are linked to those driving them, who in turn are tied to performance metrics and all that can be learned from them. Employers in the logistics industry are able to say with a high degree of accuracy where their drivers and vehicles are, and determine where they should be next. This is workplace surveillance as we have never seen it before, and it is made legible by the map, which presents a seemingly neutral picture of employees' movements.

Mentor is one such tracking technology used by subcontracted delivery drivers who are part of Amazon's North American 'Delivery Service Partner' programme, launched in 2018. It keeps track of the driver's location, driving behaviour, delivery speed and even phone calls and texts made on shift, all of which are used to pro-duce a performance score ranging from 'Poor' to 'Fantastic+'.[4] At the centre of this app, which claims to improve safety and measure performance in order to increase efficiency, is location tracking and a map that shows the whereabouts of each driver at any given time. Stuck in traffic: Amazon can see that. Diverge from the route given: Amazon can track that. Spend longer than you are required to at a delivery location: Amazon takes a note of that, too. Although the map does not determine performance scores on Mentor, it is key to how management visualizes the behaviour of its employees. After just a few years the developers of the app, eDriving, became embroiled in disputes with drivers, who claim that the technology and the company behind it disregard basic workplace rights, ignore the reality of driving on busy roads and making deliveries, and breach the worker's privacy by requiring that location services are always switched on.

The concept of the 'black box' is commonly used to describe how complex social and technological processes are obscured by the function of technology. Cars, for example, are described as black boxes because they are considered primarily as vehicles to get us from A to B, rather than complex technology built from many compo-nents, sourced from several countries, and assembled in and shipped from various parts of the world. Maps, too, can be regarded as black boxes. The complexity of the logistics, technology, politics and eco-nomics of global trade is smoothed over by the maps we use to track our purchases and sales. In effect, maps become the interface of trade and logistics because they make it legible. Most of us do not think about the processes involved in getting our items to us, but we are interested in the map that shows us where our purchases are, which is linked to the timed information about when they will arrive at their destination.

It might not seem that movement has much to do with maps. Even with the advances brought about by digital technology and smart mobile interfaces, maps are still primarily regarded as static representations, whereby places and processes are fixed securely on the page or screen. Yet maps are inextricably tied to movement, especially in the era of GPS technology. They have become one of the common interfaces that we use to view and make sense of movement, and they are intrinsically tied to many human movements, among them navigation, travel and migration. Maps are not just used to understand the movement of material goods and global trade, either. They provide an underlying graphic structure for how we understand movement over time. Consider maps that show the movement of the weather, climate change, pollution and traffic, the movements of war and insurgencies, national and international migration; those that show our exercise routes, and how far away our food delivery or taxi is; and those that help us to keep track of motorsports and lost iPhones. The list could go on.

Maps also have the potential to restrict movement, to force movement, and to free us to move. They are in this sense tied to the concept of (im)mobility, for they shape who or what can go where in any given context.[5] On the ground, the lines on the map are tied to material and invisible boundaries at both the local and global scale. Whether we are able to move across these depends on social, political, economic and environmental factors of all kinds. Maps are not the defining object that permits or prevents movement – a tall border wall, a security guard or a passport will do a much better job of that – but they do operate quietly in the background, with the power to dictate where we can and cannot go.

In this chapter, we will delve more deeply into the idea of the map as a common interface for visualizing movement, and examine why maps have been so successful in conveying a sense of people and things on the move. We start by looking at how maps were used to shape movement during the COVID-19 pandemic. We move on to examine what digital mapping technology has done to influence the perception of individual movements across the planet, and how this technology has been used to produce new forms of movement maps. Finally, we discuss the role maps play in controlling and restricting mobility over and around borders, and how the maps used to impede movement are challenged by subversive acts and natural processes.

Mapping the movements of the pandemic

Maps and movement were tied together through the COVID-19 pandemic from 2020 onwards. Perception of this virus on the move will long be shaped by the maps we saw in the news, in articles and books written about it, and throughout our social-media feeds. Where we were in relation to infection hotspots mattered, and it was maps that visualized these relations in a way we could easily understand. They could instil relief that events were happening far away, or anxiety as infection came ever closer to us or our loved ones. Pandemic maps do more than simply convey data points; they carry an authority that influences each personal perspective on the crisis.

Some of the best-known such maps are those produced by the Johns Hopkins University School of Medicine in Baltimore, where researchers collected map data about infections, deaths and recovery from medical organizations all over the world. These maps, which were used by media outlets across the globe, quickly became part of the storytelling of the pandemic. At its height, the deep reds that spread across the screen against a backdrop of black and grey signalled a catastrophic threat and the struggle to contain it. The only sign of hope was the green at the edge of the map, signalling the total number of reported recoveries.

Johns Hopkins University COVID-19 mapping dashboard, 12 May 2020.

Preceding the sophistication of these digital maps, hand-drawn epidemiological maps designed to track the movement of infection have been used for some time. John Snow, the English physician and founding father of modern epidemiology, used statistics and hand-drawn maps to track and trace the spread of a cholera out-break in central London in 1854. By mapping public water pumps, visiting each house and counting those infected at the centre of the outbreak – in what is now an altogether trendier Soho – Snow was able to produce maps that visualized the spread of the disease. His maps revealed the source of infection – a pump contaminated with sewage, now memorialized outside the John Snow pub at the intersection of Broadwick and Lexington streets – by showing that houses with high infection rates were near the pump and there-fore most likely to use it as their primary water source. Snow's maps were subsequently used to prove where the infection began, leading to the water pump being turned off, which in turn led to a decline in infections. After some controversy within govern-ment over whether the public could stomach the fact, these maps eventually ushered in the consensus that cholera was transmitted through the faecal–oral route, having previously been thought to be an airborne disease.

As is often the case with the prominent figures who shape history, questions remain over whether Snow did this all by himself. Many now believe it took the efforts of numerous scientists, statisticians and city authorities to achieve a scientific consensus on cholera and how to limit its spread.[6] Nevertheless, ever since Snow brought into the world these charts – which were legitimized among the scientific community through his standing as a respected physician and statistician – maps have been a way for epidemiologists to convey the spread of disease and shape its narrative to the general public in simple terms.

There is evidence to suggest that maps have for centuries been used to shed light on the spread of microscopic diseases. In 1694

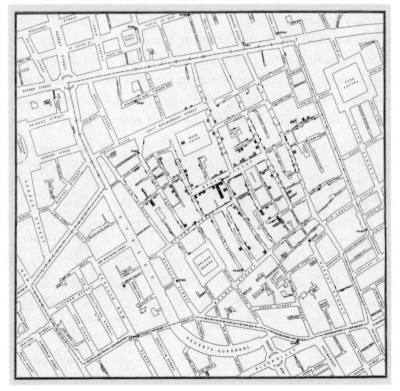

John Snow's cholera map of Soho, London, 1854.

Filippo Arrieta, a royal auditor of Italy, was charged with visualizing the spread of the plague in what is now the region of Puglia. His maps showed the areas around the city of Bari that were most infected, as well as the quarantine zone south of the city, set up by the military to curb the spread of the disease. Although not nearly as intricate or accurate as Snow's maps of London, and nowhere near as sophisticated as the Johns Hopkins digital maps, Arrieta's were nonetheless used to visualize the movement of infection and ultimately aid the decision-making of those in power.

Away from the centres of power, in the houses of those locked down during the COVID-19 pandemic, another group of cartographers was working to document the spread of the virus, how it had affected their movements and how it had turned their worlds upside down. This diverse group of amateur mapmakers were interested not so much in the science of where the virus was as in what the event was doing to the unfolding of the world. Maps drawn with pen, pencil, paper and software, put together with photographs, constructed through collage and superimposed on other maps showed how restricted movement led to a greater appreciation of

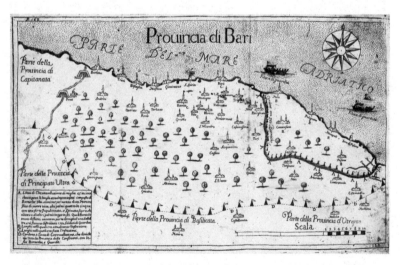

Filippo Arrieta's 1694 map of the plague of 1690–92 in the province of Bari.

Sol Perez-Martinez, *Hackney Downs Park Allotment*, 2020, mixed media collage.

Andrew Howe, *Frankwell Quay Utopia*, 2020, mixed media collage.

local areas, represented the mental stress of the situation and told the harrowing stories of the healthcare and other key workers forced to leave their homes each day.[7] Then there were those that invited us to place the mundane reality of lockdown life on to the map, in order to produce a social history of this time for generations to come, and those that encouraged viewers to think of possible ways out of the pandemic, and imagine how we might live our lives after the event.[8]

Together, these makers highlight the way maps can do more than simply show where the virus and its variants are going next, or how it might be contained. Such maps are not to scale; they break all kinds of cartographic standards, are inherently messy and intended to show relationships as much as places. But they are, nonetheless, *mappings*, because they lay out and represent the spaces and places in which lives are lived. They show that maps can be used to represent the experiential and emotional, as well as the scientific.

Mapping me

These personal cartographies buck the trend of maps made to show world trade and pandemics, which tend to be produced with large scales and the representation of population-level geography in mind. The viewer is given a universal picture of what is happening across the Earth as it appears laid out on a two-dimensional surface. However, advances in aerial photography, GPS sensors and digital mapping interfaces mean that many people are now used to tracking and mapping their own movements, too. Such maps are all about representing *individual* geography. As maps have become more mobile and location a key feature on more smartphone apps, this has become a routine part of everyday life, normalized in such practices as driving and walking.

The 'you are here' dot that places the user on the map using satellite technology started life in 1960, when the World Geodetic System was established to set international standards for global coordination. In effect, the system placed an ellipsoid – a hypothetical geometric cage – around the Earth that set a ceiling for satellite measurements. In the 1950s Europe, America, Russia and Japan had all been developing satellite programmes designed to locate aircraft and aid military targeting and navigation, but it was not until the international effort led by the u.s. Department of Defense (DOD) that accuracy in locating receivers on the ground began to improve. Years later, in June 1977, the DOD launched the Navstar Global Positioning System (GPS), which set a further benchmark in accuracy for satellite navigation systems by triangulating the position of a receiver on the ground with several satellites in space. The result was a system with the potential for sub-centimetre accuracy, which was revolutionary compared with existing systems that were accurate to between 10 and 20 metres (33–66 ft).

Until 2000 the use of Navstar was controlled by the DOD. There was a commercial signal that could be used with permission – it was, for example, adopted by the u.s. aviation industry in 1983 – but this was monitored closely by the military, which could, in theory, switch it off at any time.[9] Many uses during this period had a military focus; most famously, GPS had a major impact on operations in the 1990–91 Gulf War, when it was used extensively by u.s. forces to guide missiles.[10]

On 1 May 2000 GPS changed again when President Bill Clinton made the commercial signal available to anyone who wanted to use it. This not only took the control of GPS out of the hands of the DOD, but ushered in a new era of mapping practices, many of which remain today. Navstar has since become an eponym for GPS, and is commonly used as shorthand for the technology that is used to keep track of things, despite the availability of similar systems, such as BeiDou (launched in China in 2000) and Galileo (the European Union's recent addition, which went live in 2016).

The onset of popular commercial aviation, which unfolded in tandem with the history of global mapping technology, brought one of the first consumer uses of GPS for tracking and mapping movement. Introduced to the world by Airshow Inc. in 1982 and implemented by the airlines KLM and Swissair as early as 1984, digital maps that showed where the plane was were a way to keep passengers informed and entertained. Almost all passenger jets now have in-flight tracking maps on their entertainment systems. Similar maps have even been developed to track flights in popular simulator software. For instance, MAP! by FeelThere promises to track and display your Microsoft Flight Simulator on the map, just like the real thing.

Such maps have become folded into the normal experience of flying, even if they are not the first thing that comes to mind when one boards an aeroplane. Today, we look at them out of curiosity or as a countdown for how much longer we must sit in our seats, as something to stare at when we have watched enough movies and read enough of a book or the in-flight magazine. This does not, however, mean that we should take an uncritical view of them. As the media scholar Nitin Govil's study of in-flight entertainment systems shows, these maps blur the borders of the world and show us effortlessly crossing them, presenting us with an illusion that we are free to go where we want without the baggage of geopolitics or the reality of topography.[11] These maps do not tell us the truth. Like the commercial vision of air travel itself, they shape the perception of uninterrupted movement by glossing over the many social, political and environmental elements that have to come together to make it possible.

At the same time that they are maps to feed fantasies of travel and exploration, shaping the sense of what is possible, they also have the potential to entrench and indulge a colonizing logic that we can and must explore, dominate and oversee the lands beneath us. There is also the question of why we look at them in the first place, to which there are many answers. An eight-year-old child flying for the first time will have a very different view from a frequent flyer with a work deadline.

The idea of self-tracking has become big business in the health-and-fitness industry, and the market for wearable tracking devices is valued at $71 billion in 2023. Tracking that training session, that

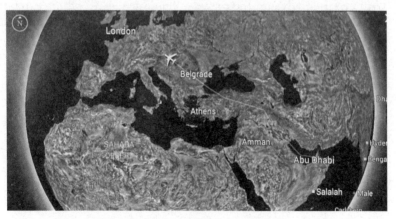

Airshow in-flight mapping interface, 2021.

cycling trip or even that tour across the country and analysing where you went, where you were quickest and where you were working the hardest is now as common for the average person as it is for professional athletes. Much of this is done through a digital mapping interface, which enables us to visualize our movements. This is linked to a much bigger social and scientific development whereby getting to know our bodies has become a practice of quantifying our actions with technology.[12] But self-tracking is not just about measurement. Such maps become meaningful as objects inscribed with personal cartography. It feels good to see where we have been; it's a sign that we have achieved something.

Some people have taken this further by using fitness-tracking devices as tools for artworks, wielding the GPS functions to inscribe pictures and words on the map through their movement across the land. 'GPS art', as it has come to be known, is growing in popularity as people realize self-tracking's potential outside purely mapping exercise for personal gain. It began long before the proliferation of the smartphone and fitness-tracking apps, when in 2000 the artist Jeremy Wood set to work recording and mapping his movements using a handheld GPS device. He still uses a similar device to produce artworks that explore personal geography and public space. He writes of the body and the GPS as a 'geodesic pencil', which can be used to express and visualize his movements. He has produced work of various kinds, including tracing his daily travel, visualizing the grounds of a university campus and even recording his lawn-mowing routes through the seasons. What I see in these images is an eerie trace of a life lived through movement, one that is stripped of all meaning beyond where Wood has been, as if he were just another package making its way from distribution centre to doorstep.

Fitness-tracking apps, such as Strava and MapMyRun, have become the dominant tools for this form of map art. Walkers, runners and cyclists have turned their exercise into cartographic artworks, some reminiscent of 1980s gaming graphics, blocky and

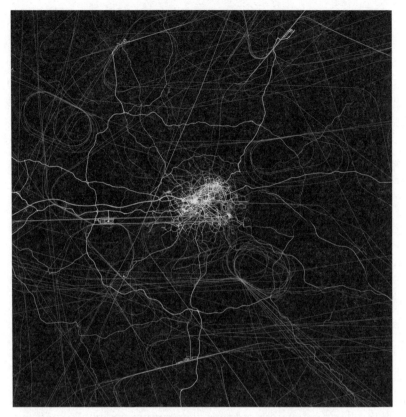

Jeremy Wood, *My Ghost*, 2016, giclée print.

pixelated. Others are cartoon-like, made all the more so by the colour palette of the base map and GPS trace.[13] Then there are those interested in experimenting with making the largest traces, or the most accurate geometric shapes possible. In 2008 the Japanese artist Yassan spent six months travelling 7,163 kilometres (4,451 mi.) to produce *Marry Me*, a real marriage proposal, which stretches all the way across Japan. To date it is the world's largest GPS artwork. Since then he has been experimenting with creating circular traces on the map by walking, which is much harder than it sounds owing to topography and the lag between GPS time stamps and the steps taken. Others have chosen to 'draw' their artwork by bike. In 2022

Daniel Rayneau-Kirkhope and Arianna Casiraghi produced a giant outline of a bicycle by completing a 6,500-kilometre (4,000 mi.) ride across seven countries in Europe, to raise awareness of what cycling can do to combat the climate emergency.

Together, these works offer more than a way to understand movement and show off the technology. They are maps etched with personal meaning and performative significance. At first glance they may mean nothing to you, but speak to the people who made them and the significance of these maps will be revealed and recognized. Maps do not exist in isolation; they are always embedded in a context that matters to someone, somewhere. In the case of these self-trackers, maps are reflections of their creators' lived, embodied experiences, and of their world view. They may not be of any significance to the mapping giants of this world – these personal cartographies are but another set of locational data points that are collected and used alongside billions of others – but they certainly matter to the people who make them.

There is another side to the maps that track movement. They are not just made for us. Such maps offer new forms of surveillance for a range of actors: family and friends, employers, total strangers, technology companies, national governments and international intelligence agencies, to name but a few. Politics are at play here, and we will discuss this much more in subsequent chapters, but it is worth noting two particular cases here with regard to self-tracking technology.

The first is a well-documented case from November 2017, when U.S. troops inadvertently gave away the location of a secret base in Helmand Province, Afghanistan, by publicly posting their jogging routes through the Strava app. Over time, the soldiers had unknowingly allowed a detailed map of the base to build up, including its location, structure, main roads, paths, entrances and exits. These so-called heat maps were a way to see bases that had been smudged off the maps of the dominant platforms, Google

Maps and Apple Maps, and created a clear opportunity for enemy forces to study the layout of the bases.[14] In the end, nothing came to pass from this exposure, but it remains an example of how even the strategy of the world's most powerful army can be undone by a group of individuals making their mark on the map.

The second case is the emergence of location-based tracking companies used by law-enforcement agencies to track suspects. One such company with reported links to the police in the United States is Hawk Analytics, which claims to offer location-based information about individuals based on their mobile-phone GPS traces, gathered from cellular operators and from tech giants, such as Google. This follows the growth in so-called geofence warrants, which are being used by police forces in the United States to gather location-based information about suspects without having to question them directly about their whereabouts on a given day.[15] This has serious implications for everyday users of self-tracking technology, because it means that the maps that purport to visualize personal details can be easily taken out of context and passed on to third parties without their knowledge. Those on the receiving end of geofence warrants are aware that this has been done only when they receive a message from the services they use, notifying them that their location data has been requested by law enforcement. In order to block its release, they are told they must fight for it in court.

Mapping the natural world

Animals have not escaped our desire to track and map movement using GPS technology, although it could be argued that such maps cannot represent their mobile lives in the same way that human trackers can. Wild and domesticated animals, including birds, bears, whales, dogs and cats, are routinely fitted with tracking devices that record their movements, migration patterns and daily routines.[16]

As in the health-and-fitness industry, there is a large market in the educational and entertainment sectors for devices and mapping platforms that claim to keep track of animals. GPS tags fitted to collars and strapped to animals, and mobile-mapping apps that purport to show their movements, can now be found in consumer-electronics shops and mobile-app stores, as well as in more specialist outlets and from makers of scientific instruments. Animal traces are visualized on maps that are used by scientists, conservationists and amateur enthusiasts alike to improve knowledge of their lives, to show the extent of their habitats and to indicate where humans may need to intervene in order to protect them. Some see this technology as a revolutionary way to predict animal behaviour, anticipate natural disasters and help to prevent the spread of infectious disease, producing data that can be used as evidence to influence public attitudes and shape environmental policymaking.[17] Others use it to shape community conservation efforts in the form of citizen-science initiatives.[18] Then there are those who use the technology to keep track of their pets, or experiment with new television formats that claim to show the hidden lives of domestic animals based on where they go.[19]

Jennifer Gabrys, a professor in media, culture and the environment, has done much to show how this mapping technology – which she places alongside a growth in environmental-sensing technology and the citizen-science movement – shapes humans' relationship with the Earth and ecological processes.[20] Her projects Citizen Sense and Smart Forests reveal how digital technology that places animals and other living things on maps alongside environmental and atmospheric data has changed the way we see ourselves in relation to other living things. She examines environmental-sensor technology in forests to track fauna and flora, GPS tagging technology to track animal migration across sky and ocean, and air-quality monitors to locate and determine the make-up of urban pollution. In doing so she points to how technology has opened our eyes to new ways of engaging with environmental and ecological processes, not to mention ushered in novel forms of ecopolitics geared towards reconnecting with the planet in the face of climate change and ecological destruction. Maps are key to this engagement because they visualize these processes and changes using an interface that most people are familiar with. Maps and other forms of 'data visualization' are especially important in this context because they do much of the heavy representational lifting in raising awareness. They are particularly useful for engaging the layperson, aspiring citizen scientist or time-poor policymaker who may not want to study large, tabulated data sets or read lengthy scientific reports.[21]

Artists have used similar technology to tell the story of this relationship. *Bear 71* is an interactive documentary that uses a virtual topographical map, motion-triggered photographs, video, audio and a series of public data sets to tell the story of how the habitat of a grizzly bear in Banff National Park was constantly degraded by human activity between 2001 and 2009. Navigating the twenty-minute documentary and the interactive map, the viewer/participant is left to piece together a harrowing picture of a life encumbered by the insatiable appetite for growth and consumption in a capitalist society. By the end of the experience, the message is very clear, and the map plays an important role in how it is delivered. We could say that it provides the visual structure for the story that is being told.

It has never been clearer that sensor technology, real-time data capture and movement maps can show what is going on in the living world, highlighting the tremendous impact humans have had on the planet. This idea is at the forefront of science and conservation work. Nevertheless, it does not necessarily mean that we have a better understanding of the Earth. Instead, it highlights the fact that we have new ways of seeing and new modes of investigating the natural world, which, as the media scholar Alison Powell has argued, potentially 'recapitulate the assumptions that human perception provides the optimal way of seeing and understanding the world'.[22] The implicit concern is whether that is enough to do anything about the loss of natural habitat and the ecological destruction that spurred these practices into action. We can map and measure all we want, to raise awareness, but the question that remains is whether doing so leads to real change in how we tackle these important problems, or whether our actions simply turn the Earth into data points – a 'programmable Earth', as Gabrys puts it – that do little more than tell us what we already know about our impact on the planet.

Moreover, the popularity of this technology further entrenches the notion that the Earth can be known only through observable and mapped data, an idea that distances us from other ways of knowing, and other ways of mapping, that could offer alternative ways of thinking about and acting on our relationship to the changing planet. There are maps of movement that take a different approach to the movements of living things, many of which existed long before so-called modern technology. These maps are not cartographic; they have no gridlines, key or systematic scale, and they do not purport to show the precise location of living beings at a given clock-time. Instead, they encourage the viewer to take a holistic look at how the movement of living things might affect that of others. These are mappings that show how humans and the Earth have lived together for millennia, and offer a way to think through this relationship in light of the current crises.

Many of these mappings can be found by looking to communities that are still close to the Earth – largely agricultural ones – where locally situated knowledge about the environment has been passed down from generation to generation through folk tales, storytelling, song, chant, sayings and tacit knowledge that can be developed only by getting your hands dirty.[23] This knowledge is inherently spatial and should be taken seriously as a form of mapping because it is used to guide such activities as planting crops in the right soil at the right time of year, and deciding which place is best suited for a particular kind of grazing. It is also spatial knowledge, closely related to care and responsibility for the land. Those using these mappings take their obligation to care for the Earth seriously, and do their utmost to ensure that this ethics of care is passed down to the next generation. In many cases this is essential for ensuring that the land can continue to return the favour by providing fertile grounds for food to be grown. It would be wrong to consider this knowledge mythical or stuck in the past, for that has been a tactic of scientific mapping: to legitimize certain forms of spatial knowledge over others.[24]

This is not a call for a return to 'the old ways'. This knowledge is not simply historical, especially for those in remote areas. Even in agri-communities that have scaled-up massively under the industrialization of farming, small traces of this remain. It is circular knowledge that looks to the past to inform the present and the future, since, when it comes to knowing the land and how to cultivate it or protect it for certain species, the past is in the future and the future in the past. These are mappings of Earth time, rather than those based on pinpoint locations and atomic clock-time, where the movement of one species is shown in relation to a wider ecology. This is the key difference between these maps and those based on GPS tracking, which tend to identify, locate, isolate and follow a single species without considering it in relation to other earthly beings and processes. For decades, many in the scientific community have seen the value in mapping the relational movements of the natural world, but such efforts are still overshadowed by the desire to map the Earth as if it were a static object, using lines and grids to portion off sections without regard or care for how the Earth moves underneath.

Maps, movement and restrictions

While maps are interfaces that show and enable movement, they are also interfaces that can restrict it. The lines on a map often correspond to borders and boundaries that cannot be crossed on legal, cultural or political grounds. These maps are not only grand geopolitical statements about who or what is allowed to move where and when, but devices that affect the everyday lives of those living in such borderlands. In these places, being on the *right* side of the line matters.

Take the line that separates the mainland United States and Mexico on maps of North America. This is no arbitrary line. Instead, it shapes an entire geopolitical narrative concerning migration and immigration that runs deep in U.S. and Mexican politics. In recent years it has gained significance in the politics of migration across Central America as so-called migrant caravans, seeking to escape poverty, violence and the impact of climate change, have made their way to the border from Honduras, Guatemala and El Salvador. The line is a popular topic of conversation and debate in Congress (in both Washington, DC, and Mexico City), on television shows, on the campaign trail and in the street. What constitutes legitimate immigration has a huge impact on migrant communities and international discourse, much of which dates back well beyond its inflammatory treatment by Donald Trump while on his presidential campaign trail in 2016.

Nevertheless, this graphic signifier, no more than a mere line on most maps, is barely questioned by those embroiled in these debates. The map and its border lines are taken as a given, the interface on which the politics have been placed. Except for a few key points along the line, such as official border crossings, the complex geography that is found along it is disregarded in favour of a binary distinction between the territories of the United States and Mexico. Effectively, the line is used simply to demarcate two sovereign spaces, but it does not take account of the land on which it is laid. On the ground this line cuts across an array of deeply significant cultural geographies and interconnected ecologies.

The Tohono O'odham Nation reservation in the Sonoran Desert, for instance, is home to a group of Native Americans. These people are no strangers to the impact of lines drawn on the map. The land on which this community has lived and worked for centuries has been cut and shaped by both the United States–Mexico national border, a direct result of the Treaty of Mesilla (1854), and the more recently created borders established in the name of conservation by the U.S. National Park Service.[25] Until 2017, the national border had been marked on the ground by a low security fence that could be legally crossed by citizens of the reservation. Residents called it a scar on the landscape and a constant reminder of how the state demarcates, cuts across and erases this culturally significant land. For years the O'odham people have had to negotiate this barrier in their daily movements, passing across it to see family and friends, visit sacred sites and burial grounds, and get to work.

In 2017 the situation worsened significantly when President Trump enacted his promise to build a border wall between the United States and Mexico. For the O'odham reservation, this meant replacing the low fence with a significant metal structure that contained few crossing points for residents. By 2021, when the newly elected President Joe Biden gave the executive order to block funding for the wall, large parts of O'odham land had already been separated by the structure. This had consequences for the culture of the community and the ecology of the land, which has little regard for artificial borders. Family and friends were now forced further apart, sites of significant cultural importance became harder to reach or were destroyed during construction work, and farmers were forced to rearrange their lands around the wall, including finding new crossing points for grazing animals and new routes for water channels. Quitobaquito Springs, a sacred water source used by the O'odham people for centuries, was hit particularly hard by these changes, and much of it dried up in early 2020 as a result of contractors sucking up groundwater reserves to mix with the cement for the wall. Estimates suggest that up to forty species of migratory bird have not returned to the springs since. Verlon Jose, the former vice-chairman of the Tohono O'odham Nation, put his feelings towards the wall this way: 'It feels like if someone got a knife and dragged it across my heart.'[26]

President Biden's executive order has put a halt to the construction of the wall, but there is little evidence yet that he is willing to take it down and reverse the cultural and ecological destruction it has caused. At the time of writing, grass-roots activism that grew out of Trump's wall campaign is gaining traction and may lead to changes on the ground, but there is still a long way to go before the wall is removed and the land handed back to its rightful owners. Even with Trump's wall gone, there remains the problem of the national border line, which would prevent a proper repatriation of the land, as it has done since 1854.

Of course, maps are not the only thing that shape the perception of migration in this part of the world – there are many factors to consider – but they have an important role to play because they are used literally and graphically to set the boundaries for the politics of movement. Jim Enote, a Zuni artist and Indigenous cartographer in the neighbouring state of New Mexico, notes that 'more lands have been lost to Native peoples probably through mapping than through physical conflict.'[27] The lines drawn on maps of all kinds are powerful because they can be used to settle claims to territory and legitimize who and what has access to them. The so-called power of maps is once again at play here, both on a grand geopolitical scale and as lived through the everyday realities of people on the ground. As the example of the O'odham people shows, however, the lines drawn on the map do not always respect the ways of the land or its cultural value.

The writer and activist Candace Fujikane reminds us that maps, and especially state-made maps, do not respect the relationality of what happens on the ground.[28] By this she means that human practices and natural processes are not easily contained by the lines on the map; they move beneath the cartographic structures placed over them. When maps do lead to material boundaries and these movements are impeded, there can be a serious impact on regional ecology and cultural practice. Water flows can be diverted, leaving farmland without irrigation; roaming animal grazers are left without enough space; and the land the livestock eat from does not have time to recover. What is more, by using maps to make boundaries, we are preventing the generative possibilities of letting socioecological processes move across and between human-made boundaries. As Fujikane's work on counter-mapping in Hawaii illustrates, cultural and ecological processes can heal and flourish in abundance when we work directly with the land and what it has to offer, rather than with the lines drawn on the map.

The border line drawn on the maps of North America has a significant impact on the movement of people. This is true in other parts of the world, too. The lines representing the border walls that separate Israelis and Palestinians, or Turks and Greeks, or those that once separated East and West Berliners, are extremely powerful in controlling and containing movement. This is because they have a material presence in the form of metal and concrete structures, and are guarded or patrolled by armed personnel. The line drawn on the map represents the control of movement, and this has a significant psychological impact, but it is ultimately the walls that prevent border crossings. Since the late twentieth century walls and physical barriers have been supplemented by digital bordering practices built from harvesting the data profiles of individuals. Today, movement is permitted and/or restricted long before we leave the house to cross the border. Transportation ticketing and international visa systems make decisions about whether we can cross the lines on the map, based on our data profiles: who we are, where we have been, whom we talk to and, increasingly, what we post on social media. This is governance and surveillance from a distance, with the help of networked technology, and has become a very popular means of controlling and containing the movement of people across the map.[29]

Nonetheless, this has not stopped people from subverting the lines on the map or climbing the walls in front of them. The spectacular scenes at Checkpoint Charlie on the night of 9 November 1989 made this very clear to the world, as thousands of East Berliners were permitted to cross the city division marked on the map. Other subversions and crossings have been less well publicized, and for good reason. Tunnelling under borders has long been effective in getting people and goods under the line, but it is often only after a tunnel has been discovered by the border authorities that this is made known to the world.

The European migrant crisis that peaked in 2015 led to a resurgence of border tunnelling in the Balkans as people sought safe passage from the Middle East. One such border was that separating Hungary and Serbia, where numerous tunnels have been discovered on the outskirts of the town of Ásotthalom. The latest was found in December 2020, when a tunnel 34 metres (112 ft) long and 6 metres (20 ft) deep was discovered, containing 44 migrants and a Serbian trafficker.[30] In response, the Hungarian authorities – keen to clamp down on migrants entering the country – have quietly constructed a fence 10 kilometres (6 mi.) long that is buried deep in the ground as well as rising from it. This has the effect of sinking the line drawn on the map well into the Earth's surface, to control the movement of people further.

Layla Curtis is an artist who works to provoke a different kind of interaction with the lines on the map. In her work *Trespass* (2015) she actively encourages participants to subvert and cross the territory drawn on the map of Freeman's Wood, on the outskirts of Lancaster in northern England. *Trespass*, a mobile app built using geofencing – a technology that restricts the app's function unless it is used within a designated area – tells the stories of peoples' relationships with the woodland, which has recently been fenced off by its landowner. Participants must cross through the fenced-off land and over the line on the map to listen to these stories, which take the form of pre-recorded interviews carried out by Curtis and loaded on to the app. Although this is an entirely different context from that of the struggles of the O'odham people or international migrants, *Trespass* shows how lines on the map – which are designed to control, contain and exclude movement – can be subverted and overcome. Even with advances in digital mapping, surveillance and bordering technology, these practices are not likely to go away, because lines on the map are never as solid as they might seem. And yet, clearly subversive actions are easier for some than for others.

*

We might be forgiven for thinking maps are useful because they show us where movement has taken place. Movement maps are sold on the premise that if we can *see* where movements have taken place, we can *know* what those movements were about. In a way, this is true. When we look at the map, we know that our package is on its way to the door, at what time our flight is likely to land, or which areas are experiencing the fastest deforestation. But, as we've seen, such maps rarely give a full understanding of movement. They are merely an interface that shapes perception. This chapter highlighted how maps of movement hide the complexities of global trade, gloss over the geopolitics on the ground and smooth over topographical reality. When we use movement maps as evidence to make claims about the world, or about changes in it, it is important to think beyond how movements are represented on the map.

At the same time, maps of movement can have real meaning that matters to people. Maps are interfaces for controlling and containing movement, but they are nothing without the politics and policy that support them. The real question here concerns who is able to move freely, and who is not. This is not decided by cartographers, but by those instructing and using them. Some people have little regard for how the lines on the map might represent restrictions on their movement, while others pay close attention to the way these lines might shape their journeys, the places they can visit and whom they can see. The power of maps, therefore, lies both in the hands of the powerful and within the representational properties of the map. Together they work to exert power over the powerless.

3

Mapping Power and Politics

In 1812 the governor of Massachusetts and a Founding Father of the United States constitution, Elbridge Gerry, signed a bill that saw the existing South Essex electoral districts remapped along partisan lines. This was intended to squeeze opposition votes from several districts into single zones in order to win elections in the 'first past the post' system, where results are based on achieving victories in electoral districts rather than on total voting numbers. By remapping this corner of Massachusetts, Gerry was able to create more districts where his Democratic-Republican party held strong, and fewer districts for the opposing Federalists, so that it was far more likely that his party would win that year's presidential election. So controversial was this tactic – a clear abuse of his power to influence election results – that the practice of remapping political boundaries became widely known as 'gerrymandering'. In March of that year, the *Boston Gazette* immortalized the phrase by publishing Gerry's map emblazoned with a salamander, to represent his reptilian and underhanded tactics.

'Gerrymandering' has since become a popular means for politicians and world leaders to draw borders and boundaries on maps in ways that favour their own interests and push their own agendas, usually for shoring up election results and winning ideological battles. For example, in Hong Kong, where electoral boundaries are redrawn at the neighbourhood scale approximately every four years,

'The Gerry-Mander', *Boston Gazette*, 26 March 1812.

it is thought that gerrymandering has been a common practice since the turn of the millennium. Research has shown that these boundary changes have clearly affected the spatial voting patterns in the country and influenced election results. Although redistricting practices themselves are not suspect – they are necessary adjustments resulting from changes in population – the way they have split Hong Kong's neighbourhoods down the middle most certainly is.[1] In some cases, it appears that entire neighbourhoods have been split based on historical voting patterns rather than population change.

How and why maps shape politics is often a question about their power. The idea of the power of maps, first proliferated by scholars in the 1980s, holds that maps reflect the interests of those in powerful positions – governments, organizations or individuals – rather than the neutral reality they claim to represent.[2] Those who study maps have long known them as tools of authority, surveillance, planning and colonization. As we have seen, maps certainly have the capacity to reshape human and physical geographies when placed in the hands of the powerful.

There is some way to go before the power of maps is taught widely in schools and reflected upon in the studios of professional cartographers, but the argument that maps are powerful social constructions is no longer the novel statement it once was.[3] Today, the power of maps is a common talking point in popular books, articles, exhibitions and television shows about maps.[4] It is a theme that is perpetuated through these media, especially when key geographical events are explained with maps, for instance to show audiences what is happening as global conflict unfolds. This is certainly the case with the war in Ukraine, where there is much amateur and professional analysis of how maps are used to wield power and influence.

At the time these arguments were first aired, however, it was a very real challenge to the way cartographers regarded their work and how the public at large looked at maps, as objective reflections of reality. Much of this work exposes the ways that powerful interests in society are reflected in maps. By 'deconstructing the map', which means picking apart the features on maps to show how they lead the gaze of the reader, researchers have demonstrated how power is both placed on the map by cartographers and harnessed by those instructing or employing them.[5] This includes questioning map projections, why it was that maps made by European cartographers placed Europe at the centre of the world, and why American cartographers did the same with the United States.[6] At a smaller scale, it means studying the place names and symbols on road and street maps to highlight the interests they reflect, and why.

A case in point is the way that symbols on Ordnance Survey (os) maps reflect specific forms of British culture that not all British people would relate to. os maps, which are famous in the UK as the standard for mapmaking, do an amazing job at representing the topography of the kingdom in great detail, but to this day they reflect a form of British culture that has long since passed, especially in urban centres. It is striking, for example, to note how all places of worship are represented by the principal symbol of Christian faith – the cross – no matter what religion is operating there, or the ways that the blue tourist-and-leisure symbols reflect a very narrow spectrum of activities.[7] Where are the mosques, Islamic centres and Hindu temples that are so clearly visible from the street? Where are the community centres, markets, social hubs and nightclubs where people spend their free time? Although os maps are beginning to adapt to life in the present day, it remains difficult to argue that these renowned maps reflect the reality of the UK's cultural diversity today. Power in this case can be seen in the ways that cartographers include some elements of British culture and not others, and in what the os regards its role to be as the national mapmaker.

Hundreds of studies have picked apart all kinds of paper and digital map in this way, from road, rail and street maps to urban plans, sea charts, strip maps, rock carvings and much more. There is good reason for this; the view from nowhere that maps claim to show has tricked us into thinking they offer an objective view of the world, when in fact they never have. This follows the philosopher Donna Haraway's famous assertion that scientific claims, which include maps, are presented as if they are universal truths, rather than 'situated knowledges' produced by people (in this case, mapmakers) who can only ever offer a partial view of the world.[8] As the process of deconstructing maps demonstrates, what maps show cannot be taken for granted, because they are *always* making a statement about the world from a certain perspective. The process of mapmaking is based on the premise that some things are included and others excluded. To include everything on a map would be impossible and render it meaningless for most uses. The expertise of the cartographer lies in their ability to make selective representations of the world, not to reproduce it. This was recognized in a short tale by the Argentine writer Jorge Luis Borges, of an empire that attempted to make a map at exactly the same scale as the territory, only to find that this would ultimately obscure the land on which it was based and render it useless.[9]

The more important question these studies ask us to consider is: who decides what to include and exclude, and what does that tell us about the world view of those who produce a map? A growing body of work has focused on deconstructing the work of cartographers, both amateur and professional, to understand the interests and actions of the people and organizations involved in mapmaking. This research goes beyond the map itself, to look at how, why and where certain maps are made, by whom exactly, and the politics and cultures inherent in making maps.[10]

In my own research with an amateur mapping community, I have found there to be many reasons why someone would want to make a map. Some mapped for fun and leisure, to build friendships with like-minded people. Others set out to be the first to map their local area, or to give something back to the community.[11] This was in stark contrast to the community of professional cartographers I worked with at around the same time. They were motivated to map for quite different reasons: to produce the most accurate map to the highest standard; to make a living; or just because they had ended up doing it after graduating from school or university. What was clear from my time with both these sets of mappers was that the power of maps did not reside only in the map itself, but rather was also distributed among the people who had produced it. This could be seen in how the mappers were instructed by organizers and managers to perform specific kinds of cartographic work within a given time frame, or the ways that power was exerted subtly by the most technically competent in the group and thus taken away from those who were less confident in their own abilities.

Labelled 'critical cartography', the work to deconstruct maps and mappers is not aiming to outlaw maps, nor is it an attack on the profession of cartography, as it is sometimes seen. It is a provocation that asks us to think about what it means to make a map, and how power flows through the maps we make. It has been, and continues to be, valuable work. However, comparatively little research on maps has focused on a wider analysis of power, as dynamic and operating at different scales and temporalities.[12] As many others have argued, power is not restricted to one dominant group, space, time or object, despite the way that narratives of power play out in society as always being exerted in a linear fashion from the top down. Studying the power of maps should therefore follow a similar logic, whereby a map's power is never entirely fixed on to the map, in time and space, or associated with the interests of those who made it. Instead, the power of a map should be understood as dynamic and situated in specific contexts of mapmaking and map use, and by examining the ways maps circulate in society. Using this framework, the power of maps might emerge differently across time and space. By examining maps in real time, on reflection and through projection, the power of maps to shape the world could be understood as a process that is in constant flux.

Chris Perkins, chair of the International Cartographic Association Maps and Society Research Commission in 2007–15, draws attention to this dynamic by highlighting the narrow lens through which the power of maps is commonly seen. I like his theory because it invites us to consider the power of maps in multifarious ways, from the power of its representation and how that may shape society, to the power these objects can have in a variety of everyday contexts:

Maps may reassure the lost, encourage debate, support arguments, keep the rain off, fire the imagination, help win or lose elections, sell products, win wars, catch criminals: an endless list of uses becomes possible, limited only by the imagination of its author: motivations may well be beyond science, even if most researchers investigating map use remain constrained by realist notions of scientific progress.[13]

In this chapter I will use two cases to highlight some of the ways we might approach the power of maps in flux, first by focusing on Google Maps, and how and why it is made and used, and second by looking at migration maps and their impact during the so-called European migrant crisis that peaked in 2015. Together, these cases reveal the dynamic ways in which power flows through maps as they are produced, used and circulated in society by people and organizations with different interests and agendas.

The power of Google Maps

Following the most popular maps of our day, Google Maps, as it is made, used and circulated throughout society, offers a way to see how the power of maps moves across and between maps, people, places, organizations and institutions.[14] This gives a sense of the various ways in which these maps have shaped society.

With a billion monthly users around the world, Google Maps is by far the most popular series of maps ever to exist. Since its roll-out in 2005, it has brought about profound changes to the ways maps are made and used, how maps are integrated into everyday life, and how value is derived from a map. Although it was not invented but procured and developed by Google, the company has found ways to maintain its dominance by implementing such mapping technology as the location dot, place search, layers, turn-by-turn navigation, traffic flow and street view more successfully than any competitor.[15] This keeps people coming back and reaffirms its influence on the way the world is represented to its citizens. Most importantly for this chapter, its popularity and influence have changed the way we might approach the power of maps. This is because Google Maps has not been designed for a specific task, as were many maps before it, but rather is designed to attract as many users as possible and appeal to as many uses as possible. These are also the first globally popular maps to run on the Internet, and now mobile devices, a fact that marked a shift in who can access the maps, where they can be used and how often they are updated. All this means that Google Maps has far greater reach in society than any other map in history, which gives us new ways to study the power of maps.

First and foremost, Google Maps represents the interests of Google, a commercially driven technology company, by foregrounding spaces and places of consumption. This can be seen when typing a place name into the search bar. The result will almost always be a map that prominently features locations where you might want to spend time and, importantly, money. Try it for yourself: search for a place and see what appears as you go deeper through the layers of the map. In 2019 Google introduced branded pins, which make this commodification of mapped space even clearer.[16] In contrast, you are unlikely to find places that are not so easily monetized; anyone who has used these maps to find rural footpaths, community allotments or local bike trails will have noted how the inclusion and accuracy of such features is tenuous at best. This makes Google Maps economically powerful because users tend to click on these places and generate data, which is analysed along with the other data users provide through Google's products and services, then sold in the form of market analytics and advertising space to brands and marketing companies that wish to target specific consumer groups.[17] The power of maps has long been tied to commercial interests, through sales, advertisements, sponsorship deals and the like, but this use of Google Maps marks a change in the ways that market value – and market power – is generated through the use of maps.

This leads us to the power Google Maps wields through the data it collects about its users. Location-based data acquired from users across Google's mapping suite (Maps, Street View, Navigation and so on) is extremely valuable, especially when it is used alongside other forms of data to develop a picture of users' behaviour.[18] Knowing where someone goes, how they get there, at what time of day and how frequently can reveal a lot about them. Google occupies a powerful position because this data is valuable to a diverse range of interests beyond businesses and advertisers. Local and national governments and organizations around the world have already begun to use Google Maps as a de facto policymaking mapping platform, which gives the platform power beyond commercial markets by influencing how these groups see the world and make decisions.[19]

Take the COVID-19 Community Mobility Reports produced during the lockdowns of 2020 and 2021. By tracing location data from its users, Google produced reports pertaining to various countries and cities, showing how mobility had been affected by the measures. The details are a fascinating and useful insight into how people's movements changed during the course of the pandemic, but this should not distract us from Google's position as the holder of this information, which is valuable to governments, planners and businesses alike.

In the consumer market, Google Maps continues to enter new territories and add new features to attract different kinds of everyday use. This is a strategy to increase the amount of data it collects from users. It has already become *the* mapping platform that most businesses and consumers use to represent and navigate the world, providing the underlying mapping infrastructure for many small and large companies. Part of this strategy has been the development of Google Maps Platform, which serves as the API (Application Programming Interface) for commercial customers. The popular ridesharing app Uber pays a hefty price – reportedly in the region of $58 million for a three-year deal – to license Google Maps to power its routing algorithm and provide a mapping interface for its users (although it is actively trying to change this by developing its own mapping platform). Part of this deal means that Google can collect data from Uber users as well as its own, offering it a window into ridesharing mobility patterns as well as the locations of its users. This adds extra value to Google's spatial data set, whereby analysis and insight of movement data can be monetized, shared or sold to a wider range of interested parties.

All this involves more than just maps. Spatial data and the computing power to analyse it are key assets of the platform.[20] Maps are just the surface layer of a stack of resources that are used to maintain Google's powerful position. Take Google Street View, for instance, which stitches together photographs to paint a picture of what a place looks like. This is not just a handy tool for wayfinding; it is a database of images that Google can analyse using machine learning to locate features to be mapped and eventually monetized. This might include details of local businesses, the specific locations of taxi pick-up and drop-off points, information about how a neighbourhood is changing over time, or even details of street furniture and road layout. By increasing the amount of granular detail on the map in this way, Google produces a map – effectively a spatial database – that encourages and attracts wider use, ultimately making it more attractive to a wider group of customers. Location-based data is a highly valuable commodity in today's economy, and Google's suite of mapping tools continues to be in a powerful position to take advantage of this.

Aside from but nonetheless connected to the commercial dominance of the platform is the geopolitical power that can be attributed to Google through its maps. By solidifying itself as the most popular commercial mapping platform, Google has become inherently powerful for showing the world where it understands certain borders and boundaries to be, how places should be named, and what should and should not feature on the map. In this respect, it tends to favour the political interests and views of the United States, and especially of non-Native American people. This can be seen in Google's maps of the United States, which show only the borders and place names that are recognized by Washington, DC, rather than those recognized by Indigenous peoples. For example, the Ohlone peoples' toponyms for the lands on which San Francisco and Silicon Valley lie, and the many sacred sites that are important to their culture, are not represented on Google Maps. Of the hundreds of examples to choose from, this seems the most apt.[21] Washington's interests can also be seen in the fact that Google Maps does not recognize Taiwan as part of Chinese territory, contrary to the view of the Chinese Communist Party; or that it has never used the word 'Palestine' to demarcate Palestinian territory, instead opting for 'the West Bank' and 'the Gaza Strip', which even so have a history of periodically disappearing from the map.[22]

These views of the world, whether you agree with them or not, are reproduced through the maps that Google makes, which end up on our phones and computers. Every time we need to search for a place, no matter what the reason, these interests are reflected back to us and have the power to shape our own view of the world. What's more, your location, or more precisely your machine's unique identifier and Internet Protocol (IP) address, determines which iteration of Google Maps you see. Take for example the disputed Ukrainian territory of the Crimean Peninsula, which when searched for on Google Maps outside Russia appears as a separate place, cut off from the mainland by a dotted line. In contrast, when searching with a Russian IP address, it is shown as part of Russia.[23] Google has come under scrutiny for mapping the world in ways that suit its commercial interests as much as its geopolitical views, and such decisions as the appeasement of the Kremlin's view of Crimea are seen as a way to avoid friction and keep the business running. Apple Maps, on the other hand, which until February 2022 had taken a similar approach, subsequently relabelled Crimea as part of Ukraine. This can be seen as either a quiet show of solidarity, or a cynical marketing ploy to be seen as being on the right side of history – guarding itself from envisaged future criticism and distancing itself from its greatest mapping rival.[24]

Power, then, is both within Google Maps and in how the maps are used to make powerful claims about the world. As Google continues to dominate the market, its maps are increasingly relied on by people from various walks of life to make claims about the 'correct' way to see the world. The power of maps then shifts from the map into the power structures of where it is used. This is affecting the geopolitics of local places as much as it is those on the international stage.

Google's Ta no Mapa (It's on the Map) initiative in 2013–16 to map favelas in Rio de Janeiro, Brazil, is a case in point. It was rolled out in partnership with AfroReggae, a not-for-profit cultural organization, to legitimize favelas in the eyes of the city and the state by putting points of interest on the map, primarily more than 10,000 local businesses across 26 settlements. Supporters of the project point to how being on the map makes a place visible and provides economic opportunities that lead to better standards of living. Being on *this* map is especially powerful because it provides exposure to a local and potentially global audience who might come to spend their money. Nevertheless, as the geographers Andrés Luque-Ayala and Flávia Neves Maia demonstrate, what at first appears to be a philanthropic mission to bring prosperity to informal settlements draws attention away from the power Google has in this context to legitimize and act as gatekeeper to informal settlements.[25] Google's maps of favelas may tie these places to the city at large, but they do so using economic logic that favours using maps to show *commercial* points of interest, rather than the many other locations that could be mapped to bridge the social, political and environmental divides from the wider city. Luque-Ayala and Maia note that 'through digitally mapping favelas, Google Maps is advancing a project where equity and social justice is [*sic*] achieved not so much through social inclusion and service provision, but rather through a calculative incorporation to a specific economic regime.'[26]

If Google wanted to make a meaningful intervention that would make favela life visible to the masses and do more than serve the company's commercial reasons for mapmaking, it could look to the participatory mapping projects in the city that use the same tools to work with favela communities to make maps that represent social and political issues, which are then used in the ongoing fight for recognition, resources, social services and infrastructure. It could take the lead from a community project that uses its maps to make a different kind of favela life visible. The Youth Forum's Militarization of the Favelas project has used participatory workshops and a dedicated app to geolocate cases of violence by the city's military police towards young Black people in fifteen of Rio's favelas since 2014, and to map youth spaces in these communities. This project seeks to make visible violent and intimidating policing. It acts as a form of youth-led resistance by producing a map that represents spaces that are significant to young people, including safe meeting points, such as football fields and courtyards, and spaces to be wary of, such as drug-dealing spots and areas where they are likely to be stopped and searched by police. The resulting map looks like a standard Google Map, but the red pins on it tell a very different story of life in favelas from those produced by the Ta no Mapa initiative. Here is a map that uses the power of Google Maps in a wholly different way, to tell a social story about a marginalized youth community and to provide information about safe spaces.

Google does have a history of participatory mapping. Between 2008 and 2017 anyone could use the Google Map Maker platform to submit suggestions of features to add to the map, a service that has since been integrated into the main mapping platform. This means that infrastructural, topographical, commercial, historical and civic points of interest can be added to the map by anyone after validation by a team at Google. In its early days, this was lauded as an example of democratic mapping done at scale, and it became a popular activity for those living in under-mapped areas, where it was used to make places visible to a global audience. In recent years this has spilled out into Street View, meaning that anyone can contribute photographs of places to be added to the database and possibly used in the public version. The latter has become especially popular for places with little Street View coverage, such as much of Africa and Central Asia. Hover the Street View icon over the world view of Google Maps and you will see just how little of these regions have Street View integration.

At first glance, these initiatives put the power of maps into the hands of anyone who wants it. The promise is that with a few clicks and suggestions *you* can have the map updated and contribute to a global mapping project. Take a closer look, though, and it will become clear that these services do not offer an equal share of the power. Contributors may make all the suggestions they want, but Google has the final say on what can and cannot be added to the map. What is deemed important to map for Google – does it have economic potential? – may be quite different from what individual contributors prioritize. What's more, contributors are not paid and are thus at odds with Google, which turns their contributions into economic capital. As with all Google's products, mapping is free at the point of use because people offer their time and data in return.[27] Google will remain a powerful gatekeeper to what can and cannot be mapped unless its business model is changed.

This analysis of Google Maps tells us that it is very easy to *say* that the power of maps is distributed rather than fixed on the map itself, but less easy to pinpoint exactly where or with whom the power that surrounds maps is located at any one time. This is because the dynamics of power are not easily captured and contained in a general analysis of maps. Google Maps is ubiquitous, has many uses and is produced by large teams of people as well as lay contributors, all of which makes it even harder to pinpoint exactly where our analysis should be focused. The contexts in which Google Maps are used today can vary from pop-quiz arguments about where countries and capital cities are to government-level decisions on whether to intervene in disputed territories. There is no one way to understand the power of a particular map, including the most popular of our time. It is therefore important to direct our attention not to specific maps per se, but to specific cases in which maps have been used, to highlight the various ways that power flows through maps and their contexts of use.

Migrants and maps

We can gain some insight into the dynamics of power by focusing on a specific case, the European migrant crisis that peaked in 2015, in which maps were used to shape public discussion and the experiences of migrants themselves, and to reflect on the events that unfolded. Here we focus on a number of maps produced in this period, from media and campaigning maps to navigational maps and those made by artists and migrants themselves, to show how a diverse range of cartographic techniques was deployed to wield power for those who made them, and how maps made by the powerful can also be used in powerful ways by migrants themselves.

Maps have long been used to spin a story about what is happening, even when it is not, or not to the extent claimed. The history of cartographic propaganda in Europe goes back to the Middle Ages, when T-O maps – circular maps of the world with a T representing the division between continents – were produced by skilled printers to show a simplified layout of the lands known to Europeans at the time. These maps are crude and unsophisticated by present-day standards of cartography, but were nonetheless used in a similar way to today's propaganda maps: to influence the world view of the reader. Since then, maps have been used for political and ideological propaganda of all kinds, from election campaigning to war reporting.[28] Many people will recognize the government-issued propaganda maps produced during the Second World War to reassure the public that the fight was being won, or the pastiche of these maps used in the title sequence of the popular television series *Dad's Army* (1968–77). Key to the success of these maps was the use of colour, iconography and slogans, which direct the reader's attention and instruct them on how to read the map. Propaganda maps used in wartime are known for their bright, contrasting colours to represent territory and bold, sweeping arrows to show movement. These are intentional tools deployed to win the argument the map is trying to make.

Earliest printed example of a classical
T-O map, by Günther Zainer, 1472.

In recent years maps with similar tropes have become vital tools for shaping understanding of the movement of asylum-seekers, refugees and other undocumented migrants. At the height of the so-called European migrant crisis of 2015, journalists, commentators, politicians and humanitarian organizations were all using maps to show the movement of people, where they were at particular times and from which borders they had been closed off.

That year 1,005,504 people from the Middle East and North Africa arrived at the southern borders of the European continent, seeking refuge and the chance of a better life. Beside the momentous practical challenges this presented to European authorities in keeping these people safe, housed and fed, and the equally arduous task of processing so many applications for asylum and citizenship through systems that were not properly prepared, this mass movement of people across the globe quickly became an ideological challenge that manifested as an identity crisis for many European nations. Right-leaning governments and the return of nationalist opinion-makers used the crisis to bolster the popularist message that they should reclaim their sovereignty and national identity, and therefore keep these people out.

Much of this messaging was based on the premise that photographs, maps and slogans could reduce the complexity of migration to simple campaign-driven narratives. Photographs and maps can be used to simplify complex subjects, but they can also distract from the important details. There is a seductive quality of imagery that claims to do the hard work for us, and maps fit right into this. They give us licence – certified and stamped by the authority of cartography – to say to ourselves and to others that we understand what is going on where, when really we do not. The abstract understanding offered by the map is frequently far removed from the details on the ground, a fact that should encourage us to question why we so regularly reach for the map to help us make sense of complex issues.[29] In the case of the migrant crisis, it would be wrong to suggest that the public was all completely convinced by what was represented on these maps – we have more agency than that – but I suspect that the general view depicted did relieve most people of the effort of seeking out the details of this mass migration, the nuances of migrant lives and reasons for travel. I, for one, admit that maps helped to give me a general impression of what was happening, but they were not at all useful to my understanding of the movements of people on the ground.

A clear example of how maps were used to smooth over the complexity of the migration crisis and the public rhetoric concerning immigration can be seen in the run-up to the Brexit vote of 2016, when maps were used by Leave campaigns to feed the narrative that migrants to the UK take jobs, exploit welfare systems and pollute British culture. The now infamous 'Breaking Point' posters unveiled by the UKIP leader and hard-line Brexiteer Nigel Farage used an image of migrants walking the 8 kilometres (5 mi.) from Dobova railway station to the Brežice holding camp in Slovenia to stoke fears and fuel a culture war. Shortly after their release, the posters were criticized as propaganda and compared to the tactics used by the Nazis to shape public opinion about 'other' populations.

Less discussed were the maps produced by the official Vote Leave campaign and the press that supported it, which used a range of cartographic techniques to depict the perceived effects of an uncontrolled immigration policy on migration from Turkey (a country that was seeking EU membership) and its neighbouring states Syria and Iraq. Notably, the maps used bold red arrows – known for signalling danger – to gloss over the complex geography and bureaucracy standing in the way of potential immigration from the region. This was both a reductive cartographic campaigning technique and a self-aggrandizing claim that migrants from the region would not want to settle anywhere except the United Kingdom.[30] While Farage's poster was widely vilified, including by the official Leave campaign, led by Boris Johnson, these maps largely

The EU is letting in more and more countries

- The EU started as 9 countries
  – it's now 28

- Croatia, Romania and Bulgaria
  have joined since 2007

- Five more members with a total
  population of more than
  88 million are being lined up
  to join

The EU will continue to grow.
The next countries set to join are:

Albania:	2.8 million
Macedonia:	2.1 million
Montenegro:	0.6 million
Serbia:	7.2 million
Turkey:	76.0 million

Syria   Iraq

www.voteleavetakecontrol.org

Vote Leave

Vote Leave campaign map for the UK's referendum on leaving the EU, 2016.
The map displays the fear promoted by the Leave campaign of how the expansion
of the EU would increase the number of migrants to the UK from Syria and Iraq
via Eastern Europe.

went under the radar despite there also being a well-documented history of cartographic propaganda being used in Nazi Germany.[31]

The Vote Leave campaign and its allies produced these maps and subsequently won the vote, which suggests that it was they who harnessed the power of maps to influence minds. Yet this analysis is too simple because it is formed in hindsight, after the vote was won, and ignores the many other elements that may have influenced public opinion before and during the campaign. It also assumes that the readers of these maps became, or had the potential to become, Leave voters, which distances them from the possibility that they had no interest in the maps or Brexit at all, or the possibility that they made use of these maps for their own political goals.

The Cambridge Analytica scandal of 2018 raised awareness of how Facebook was being used by the Vote Leave campaign to push personalized messaging, including maps, to potential Leave voters.[32] However, it did not reveal how Remain campaigners and others were also harnessing the power of Brexit propaganda maps on social media and beyond, for their own interests. Arguing them to be a false representation of the facts became a social-media trend for a time, with journalists, newspapers and amateur fact-checkers and bloggers putting them up against migration statistics and real-world examples to prove their message wrong. These people were reappropriating and subverting the power of these propaganda maps for their own agendas.

The same maps of migrant movement designed to provoke anger that people were coming to 'take' what was left of a dwindling job market, or push already stretched health and welfare systems beyond breaking point, were also used to highlight what was seen as lies, and provoke sadness and empathy for the lives of the migrants. So, the fact that the Vote Leave campaign was victorious, and that maps were a key part of its messaging, does not automatically mean that the power of maps resided only in the hands of Brexiteers. Power is not something stable in the hands of the powerful, but rather a force that is always in flux between people. Once these maps were released, it was up to the public how they were used and believed, and ultimately what power they had (and perhaps still have). This is not to say, however, that there are never any power imbalances when dealing with maps. Of course there are; maps in the hands of one person, institution or organization may be much more powerful than in the hands of others. But the power of maps to influence minds is not an exact science, and maps produced by those in positions of power do not always shape the views of the reader in the intended way.

*

As the map scholar Laura Lo Presti has noted, maps of the European migration crisis rarely gave the full picture of individual journeys, many of which were characterized by stops and starts, practical considerations, politics, bureaucracy and emotional experiences.[33] Instead, the maps used at the time tended to smooth over the details and provide a simplified view of events. In her assessment, more could be done to help the public understand the power of maps in shaping the perception and experiences of migrant journeys. One way she suggests doing this is to discuss maps used by migrants, their loved ones, activists, lawyers and humanitarian organizations as 'emotional mediators' that convey the lived experiences of migration. By engaging with maps in this way, using them to convey stories rather than statistics of migration, less emphasis is placed on the objective perspective of maps and more on how they can be used to understand human lives.

Focusing on a range of maps used during these events offers a different lens through which to understand the power of maps. Many maps were used to put power and agency in the hands of migrants and their allies. There are, for example, reported cases of migrants using Google Maps for navigation across borders, to bookmark and share key locations, to find a safe passage or, as others would, to find the nearest supermarket or cash machine.[34] As we have seen, Google Maps is powerful in representing the interests and world view of this large corporation, but equally it is powerful in helping migrants to navigate across land. Then there are the cases of migrants using WhatsApp and Facebook Messenger to share location details so that friends, family and trusted allies can keep track of them.[35] This is both a coping mechanism that lets others know they are on the move, and a legal and safety measure to ensure that responsibility is taken to locate people if something goes wrong. In the fatal crossing of the English Channel on 3 November 2021, in which 27 people lost their lives, the significance of the location pin came painfully to the fore with reports of families who were no longer able to track their loved ones as the news came in. Some were left with haunting maps showing their relatives' last known location out at sea.

Watch the Med is an organization that has produced an online platform to map and keep a record of migrant journeys across the Mediterranean Sea. Anyone is able to participate in the map by sharing information about migrant crossings, sea conditions, local rescue operations, activist networks and much more. By mapping this information alongside reports of individual cases, they are harnessing the power of maps to highlight where migrant deaths and human-rights violations occur. This is important information for past, present and future migrants, and puts pressure on the coastguard authorities that must take responsibility for them.

These are forms of counter-maps, made to represent that which is commonly left off maps by those with established and institutional power. In recent decades this type of map has become a successful tool for marginalized people from around the world to use to put their lives and lived experiences on the map. Maps made in this way are used alongside other documentation and representation to fight battles over land, rights and ways of life. They have been used in many contexts, from maps of displaced people, cultures and languages to maps that highlight Indigenous land rights and sacred spaces, and those that put local issues on the map, such as those produced by the Youth Forum in Rio de Janeiro.[36] We'll talk more about the power of these maps in Chapter Four.

Finally, we should consider the maps and mapping technology used by border authorities to keep track of the movements of migrants, both while they are in transit and once they have claimed asylum. These are not counter-maps, but rather mapping and tracking technology bound up in national-security infrastructure, and they perhaps represent the ultimate power that maps can have. Drones, satellite imagery, GPS tracking, biometric passports and so-called smart bordering technology that use international databases, algorithms and machine learning to identify and predict potential movements and border-crossing points are all common tools for governments around the world to keep migration under control.[37] Prediction is a key factor, and border authorities try hard to prevent migration before it happens, rather than react to it after it has occurred. This trend for enforcing borders at a distance has been made possible by digital technology, much of which uses maps to make sense of migration flows and pinpoint where action should be taken.

The power of maps in this instance lies primarily in the hands of the national governments, which have the power to use them to suit their agendas when it comes to immigration policy. But we must not forget who is producing much of this mapping technology: private companies vying for lucrative government contracts. They, too, are in a powerful position to shape how this technology is built and used, including how mapping technology is designed and used to keep immigration under control.

By turning our attention from the way the migrant crisis was represented on a single map to different maps and other uses that maps had during this time, we can see how the power of maps shifts depending on where you look, shaping public discourse on the migrant crisis, the lived experience of migrants and the control of movement across borders. Similarly, if we take Google Maps at face value and do not examine the context in which it is used, we get only a narrow view of how powerfully maps shape the world.

It is now common knowledge that maps and power cannot be separated, but the map–power relation is not straightforward. The power of maps cannot always be understood simply by studying what is on the map and what effect that can have. It is a dynamic relationship that is shaped by real-world events that unfold in situ, meaning that we can never entirely predict how the map–power relation will play out just by looking at a map. What map is being used? Who is using that map, where, and at what time? How is it being used alongside other instruments and institutions of power? These are all questions that must be answered if we are to understand the real power of a map in a given situation. It is not enough to say that the map has power all by itself. It is never just the map that changes the world, but all too often the map is analysed in isolation: *this* map is powerful because of the way it represents the world, or *this* map is powerful because of the effect it is having or has had on the world. Analysis based on isolating the map in this way provides a workable frame through which to view the map–power relation, but it does not get us any closer to understanding the instability of a relationship that is never fixed. The many uses of Google Maps and the situated uses of maps deployed during the ongoing migration crisis are exemplars of this instability.

4

Mapping Culture

Encased on the walls of Hereford Cathedral in the west of England rests the largest known map from the medieval world, inscribed on vellum by Richard de Bello and dating back to about AD 1300. This 'Mappa Mundi' looks very different from the maps that are used today; it's circular, for a start, and among the cluttered iconography of people, places, creatures and angels, there is very little that resembles the scale, order and orientation that we tend to associate with cartography. Like a great artwork, it requires some time to take it all in. And, in many respects, it *is* an artwork, not a geographical map. Experts have argued that it does not correspond to the geographical knowledge of the world at the time, which was far more accurate than the map suggests. Instead, it is a map of Christian culture, inscribed intricately with important places, people and imagery from Christian history and beliefs. In the centre lies Jerusalem, representing the birth of Christianity. At the top sits the Garden of Eden, more real and less symbolic than it is considered today. In between are spiritual places, people and animals, and the pathways that connect this religious culture.

Mapmaking has changed a lot since the Hereford Mappa Mundi was drawn. The science of mapmaking – cartography – was developed during the European Renaissance, and maps became associated less with culture and more with the physical features of

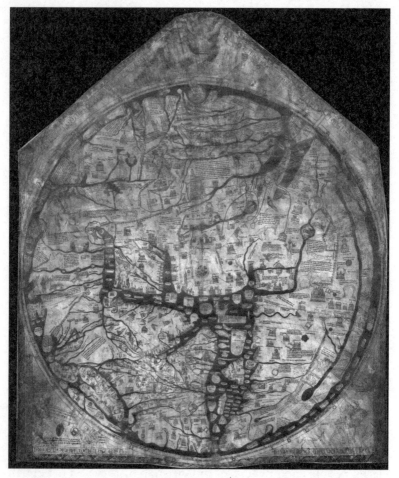

Richard de Bello, Mappa Mundi, 14th century, on display at Hereford
Cathedral, England.

geography: the roads, rivers, railways, hilly contours and human
settlements that we now expect to see on a map. Culture seemed
to disappear from the map in favour of accurate scientific representation. Maps were not made, sold and acquired based on their
ability to represent culture. They were instead largely geared towards
practical applications that required accuracy and objectivity. They
became objects of a popular scientific imaginary, and buyers

subscribed to the notion that the world could be understood from above and afar.

But a closer look at maps in the modern era shows that they do still represent and shape culture. Culture does not just get stripped from the map in favour of science. Take the famous Michelin road maps that still take up space in many a car boot, glove compartment or back seat. These iconic maps have a reputation for accurately representing road networks at the same time as they have become an object and symbol of car culture. They are both useful navigational aids developed using scientific surveying methods and emblematic of driving and all that comes with it.

First published in 1900 by André and Édouard Michelin of the well-known tyre firm in Clermont-Ferrand, France, these maps were made popular around the world through the twentieth century, alongside the rise of tourism and mass car ownership. Initially a mix of travel guide, maintenance manual and city-centre maps given away freely to promote the tyre business, these pamphlets evolved with the launch of a 56-page national atlas in 1907.[1] In France, the atlas had the obvious benefit of helping a nation of emerging car drivers navigate the road network, but it had an even greater impact by influencing the administration of roads themselves, being partly responsible for the move to road numbering by the French government in 1913.

Michelin began to produce maps for the United Kingdom (based on data from the Ordnance Survey, incidentally) in 1914, and later for many other countries around the world. After the Second World War the growth of tourism and cheaper mass-produced cars provided the perfect conditions for the maps to sell well commercially, and they became a staple companion for many drivers and navigators. Today, even with the growth of digital maps and a large market for road maps, Michelin still sells large quantities of paper atlases each year, alongside the development of its own digital journey planner. In 2021 alone, sales of the paper maps

Michelin road map of Belgium, 1940.

jumped by 20 per cent. Whether or not this is an anomaly produced by the pandemic, it goes against the narrative that we are
all satnav users today.

Like their sister products, the Michelin Guides to restaurants,
the Michelin road maps were developed as a savvy marketing strategy. The Michelin brothers did not intend their business to become
a famous mapping company, or to be associated with the final word
on the quality of dining, yet this is what they have become. Both
publications are deeply embedded in driving and food cultures all
over the world; they are not *just* guides and maps. That is quite a
feat when we consider that the original plan was simply to sell
rubber tyres.

To make a map is to make decisions about how cultures are represented and to make an object that enters and circulates through
culture and society. In this view, culture never left the map and maps
never left culture. This is as true of maps made for the cultures of
everyday life, where humans share norms, values and interests (such
as car culture), as it is for maps that represent culture in the forms
of geographical, historical, artistic and spiritual identities.

In Britain, maps are bound to the nation's cultural identity. They
shape and are shaped by what people think of as 'being British'. At
the turn of the millennium, if you asked a British person who made
the best maps, they were likely to name the Ordnance Survey, the
country's national mapping agency since 1791. Many would still say
the same today. Ask why and they might have said it was because
os maps gave the most objective and accurate view of the land, offering the example of how that hill in front of them corresponded with
the hill shown on their map. They were not wrong, but this was and
still is a cultural perspective based on a shared understanding about
what maps are designed for, how to use them and who emerged as
the dominant national mapmaker in public consciousness during
the nineteenth and twentieth centuries.[2] Even at the peak of the
os's popularity, before the dominance of Google Maps, this was

not a perspective shared by everyone, but rather a dominant idea taught and passed on through formal education and common cultural practices, such as navigation and outdoor activities, where os maps were widely used.

As Google Maps has become the go-to map for British people, the os has had to reassert itself within British culture, positioning itself not as the best map for navigational use but instead as a specialist in accurate and detailed maps for outdoor pursuits and (a fact that is lesser known) a data company that produces precise spatial data sets and models for the public and private sectors.[3] The os is certainly still responsible for shaping the British perspective on maps – os maps are still the only named maps on the National Curriculum for England and Wales for geography, for example – but there has been a shift in the ways these maps shape everyday cultural practices, a shift that has arguably been led by Google and mobile mapping more generally. This shows how the relationship between maps and culture is always changing, even when it comes to maps as entrenched in national cultural identity as those of the os.

In the digital era, we still see culture all over the map. With each zoom magnification the map tells us something new about which culture is represented and reflected, and which is not. The restaurants, museums, galleries, institutions and historical sites of interest that emerge alongside the shops, religious sites, amenities and street names all shape our understanding of place and of the people who live there. This may have an impact on whether we decide to go there, whether we consider it a place for us or not. As the geographers Mark Graham and Martin Dittus have argued, what is represented on digital maps matters for the recognition of cultures.[4] Being excluded – whether intentionally by a map's designer or systematically by algorithmic logic that favours the most popular clicks, links and languages – can make a huge difference to how and why a place is visited, perceived, understood and even governed.

Digital maps also reflect wider technology-focused culture. These services are so ubiquitous today that we can be forgiven for forgetting that they have only shaped everyday cultures of different kinds since the turn of the millennium. Commuting cultures, urban cultures and consumption cultures are just three that have been affected by the popular rise of digital maps, and especially of Google Maps. In the year 2000 paper road maps were still the norm, urban navigation was largely done with a pocket atlas or street map, and the ability to search in real time for shops or services was afforded only to those with deep pockets for the latest technology and the patience needed to use the slow mobile Internet at the time. Since then, Google Maps has become a cultural object that represents a significant shift in the relationship between maps and culture – from how cultural institutions and practices are represented on the map to how everyday cultural practices unfold.

Nonetheless, maps only ever give a partial view of culture. Culture is a slippery, ambiguous concept, especially when we are using the blunt tools of cartography in the name of science. Decisions are inevitably made by mapmakers about what cultures appear on the map, and what place names, points of interest and information to include or exclude. Such decisions are inevitably influenced by the cultural perspectives of mapmakers themselves. We will talk more about these decisions later in this chapter, but first a quick note on mapping perspectives and the culture of cartographic abstraction.

Professional cartography continues to be dominated by a select group of the well-educated middle class. We have come some way from the largely white, male teams of professional and academic cartographers that characterized much of mapmaking in the twentieth century, but we are still far from the inclusivity that is seen in other spaces, such as web-based amateur mapping and counter-mapping (even though those are not perfect). Professional cartographers and their institutions have done well to produce accurate scientific maps that show the lie of the land or get people to where they want to go, but they frequently fail to represent anything other than their own cultural perspective on the map. We know this because the maps of the national agencies and commercial enterprises that dominate professional cartography do not reflect the many ways in which different cultures view the world.

At the most basic level, the maps of professionals reproduce the dominant Western perspective that the Earth can be understood only by looking down on it from an imaginary viewpoint. This perspective developed over time as buildings increased in height and aeronautical technology and photography took off. Most professionally designed maps favour this abstraction, which has hidden itself so well behind the veil of science that few perceive that these maps represent a cultural viewpoint at all.[5] But they surely do, if we consider that maps found in other mapmaking cultures do not view the world in such abstract form. Pictorial mappings and story maps, for instance, offer other ways to view the world from a cultural perspective that favour lived and grounded experience.

Jina Lee's TalkingMaps project (2016) is an example of mapmaking that takes a different view. She mapped the migration experiences of Joseonjok women from China's Jilin province to southwest London, using paper, illustrations and text to fold the abstract view of their situation into a pictorial and written narrative about their lived experiences. Some readers will not regard these as maps at all, or will categorize them as part of a 'non-conventional' mapmaking culture, but these are maps that favour a perspective born out of an alternative belief system of how the geography of the world should be represented. They remain carefully selected spatial representations of place, just like any other map. The dominant cultures of professional cartography could learn a lot from these ways of seeing, should they be willing to think beyond the abstract view.

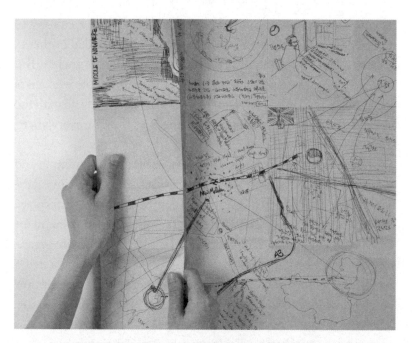

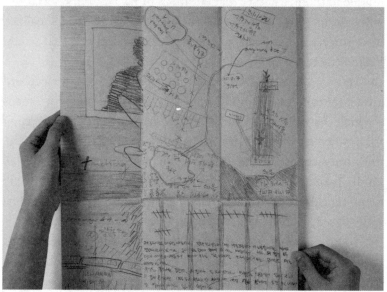

top: Jina Lee, *TalkingMap #1 (detailed)*, 2016, pencil on paper.
bottom: Jina Lee, *TalkingMap #2 (detailed)*, 2016, pen on paper.

So, maps represent culture, mapmakers select which cultures to represent (even if this is unknown to them or unintended by them), and maps become symbols of culture that circulate in society. But there is more. There are also cultures *of* the map, people who share social norms and practices that are shaped by what they do with the map. We can call these mapping cultures, and they are everywhere, if you know where to look.

Leisure pursuits like map-collecting, orienteering, gaming and travelling have all become mapping cultures, many of them established over the past two centuries. Then there are the workplace cultures of professional cartographers, the cultures of amateur mapmaking and humanitarian mapping activities. This is where organizational cultures intersect with cartophilia, with making a living and with trying to make a difference. We might also consider how we as individuals develop our own cultures *with* the maps we use. Many people will have developed a style of map-reading that is unique to them, whether that be rituals of navigation, associations with certain kinds of map or marking up maps with personalized inscriptions – significant places, names and details – that would make little sense to anyone else.[6]

These are diverse cultures that do not always intersect and attract the same people. They share a common denominator – maps – which shape how they function, but maps are not the only thing to consider. As with all cultures, mapping cultures are not bounded. They intersect and are shaped by social and cultural activity of all kinds. Take as an example the cultures of orienteering. These are mapping cultures inasmuch as they are leisure cultures, geographically based, shaped by race, gender, age, education and money.

Today mapping cultures exist online as well as in person. Social media especially has created new spaces for these cultures to develop and form new connections. The hashtags #amapaday and #mapsinthewild, and the handles of famous map libraries, including the British Library's Map Library (@BLMaps) and the David

Rumsey Map Collection (@rumseymapcenter), have brought together a diverse collection of maps and enthusiasts across platforms. The #30DayMapChallenge, started by Topi Tjukanov in 2019, challenges people to make a map based on a different theme each day in the month of November every year, and share it on social media. Tjukanov has said that the event is a way to build a global community of mapmakers. By not specifying what type of map to produce, the event attracts a diverse range of mapmakers, from professionals to first-timers, and a variety of maps, from digital renderings to maps made from foodstuffs and plasticine.[7]

These are all signs that mapping cultures are alive and well. 'I love maps!' or 'I'm fascinated by maps' are statements that I hear very often, and it is difficult to find people who admit that they hate maps. This has not gone unnoticed. There is an entire industry built around mapping cultures, ranging from the sale of collectible and personalized paper maps, to the establishment of societies, to the development of games and forms of travel that are influenced directly by the map. Mapping cultures are, of course, nothing new. People have long found a passion and a use for maps in their spare time, especially since maps became available to the masses in the nineteenth century. But with so many forms and uses of maps today, it is little surprise that maps have become a much-loved part of what people do with their lives.

For as long as people have made and used maps, culture has found its way on to the map and maps have found a way to shape cultural practices. This chapter sheds light on both, with the aim of complicating our cultural perspective on what maps are for, how they should be used and who should make them. Beginning with cultures *on* the map, I show how dominant forms of culture end up on the map and how this is being subverted by marginalized groups through counter-mapping. I then turn to cultures *of* the map to highlight how maps have become a popular object for people to collect, congregate and socialize around, and work and play with.

Cultures on the map

Mapmaking has always been about making decisions about what to highlight and what to obscure. The best maps are generally regarded as those that clearly show their intentions in a way that is both useful and aesthetically pleasing to the user. Cartographers, mapmakers, experienced navigators and wayfinders, along with cartographic departments, national mapping institutions, cartographic techniques, mapping technology and specialist societies, build a reputation based on their ability to produce maps according to these criteria. There is a whole industry and library of literature that is geared towards producing maps to satisfy these aims. There is not, however, much in the way of discussion about the conditions and contexts in which these criteria are followed and by whom maps are made. This matters, because it is people, their histories and how they work that determine what ends up on the map.

The narrow view of culture on a map can be seen by looking at it through the lens of gender, specifically how male perspectives typically end up on the map while female perspectives are often left off it. This is despite the fact that plenty of women have made maps throughout history, as both amateurs and professionals, and despite the efforts made by professional cartographers, mapmaking societies and communities to address gender inequality.[8] Since the early 2000s numerous gender-parity initiatives have been put in place by the profession of cartography and cartographic communities set up specifically for women.[9]

There are famous female cartographers, such as Phyllis Pearsall, who founded the Geographers' A–Z Map Company, producers of the famous *London A–Z* pocket atlas (first published in 1936); the pioneering oceanographic cartographer Marie Tharp, who produced the first scientific map of the ocean floor (with Bruce Heezen); and the so-called Military Mapping Maidens, two hundred women who produced maps for the U.S. military during the Second World

War. However, it is only recently that these cartographers have been recognized in a field usually dominated by men.

I am not arguing that maps *simply* represent the world view of men (a universal view of men surely does not exist), nor that women have no agency in the production of maps, but wish rather to say that the principles, practices and spaces of mapmaking (especially in the science of cartography as developed in Europe and the United States) emerged largely in patriarchal societies where the male perspective dominated, and still does. This follows the dominant Western-masculine view of geographic sciences more broadly, which views the landscape as an abstract resource for the production of scientific knowledge, rather than as a life-support system with which all humans are intertwined. The feminist geographer Doreen Massey writes that

> Gender has been deeply implicated in the construction of geography – geography as uneven development or regional variation and local specificity (and in the construction of these, not merely the fact of them), geography as an academic/intellectual discourse and set of social institutions, and geography in terms of its founding concepts and systems of knowledge. In particular – the concern here – gender is of significance to geographical constructions of space and place.[10]

With Massey's words in mind, the funding, research, management, administration and production of cartography – for the most part a distinctly geographic discipline – has tended to be controlled by men. This has an impact on the kinds of map made and on what ends up on the map. Under these conditions, even maps produced by women can represent the world from the male perspective.

To say that maps are gendered goes against the ideal that maps offer an objective view of the world. Many will disagree that gender has anything to do with maps. But a closer look reveals gendered perspectives all over the map, from the way spaces are represented to the way maps are used. Here I take the lead from the feminist writer and activist Sara Ahmed, who argues that feminist critique cannot be something that is simply applied from the outside – a criticism of stereotype that is frequently aimed at feminism – because at present there is no outside unaffected by patriarchal relations. I am not somehow searching for a feminist critique or angle on maps, but rather acknowledging that maps, as with so much else, have emerged from a patriarchal society and therefore cannot escape its influence.

Mobile maps used for walking are a useful example of the gendered dimension of maps. On the surface, they appear to be gender-neutral representations that many people use every day for wayfinding. They depict routes to and past amenities of all kinds, from churches, schools and post offices to pubs, parks and police stations, and tell us when to take a turn and how long it is until we arrive at our destination. But when we consider how the map represents these spaces and gives directions – fixed and detached from the ways they might be experienced in the real world – we begin to see a particular kind of male perspective, one based on the perception that anyone is able to roam wherever they like without consequence, regardless of context. This is a detached perspective that does not consider the gendered differences in how people get from A to B.

Routes through parks, for example, are almost always shown as devoid of any dangers that might exist. These green spaces look like islands of refuge in the urban sprawl. Similarly, footpaths, passages and alleys appear the same as any main road, as spaces that can be navigated with ease. These are probably innocent design choices made by well-meaning cartographers, but they nonetheless per-petuate the heteronormative male perspective that space is neutral.[11] The world is in reality more complicated than that, and gender plays a role in navigation. Men and women looking at the same map may interpret it differently, as space to be navigated with ease or as space to be navigated with dangers in mind, depending on their knowledge of the area and the situation they find themselves in.[12]

Studies have shown that men and women navigate in different ways. Men are more likely to use shortcuts and an efficient orientation strategy, while women are more likely to wander and use known routes in a situational manner.[13] While this research is interesting, it would be wrong to suggest that navigation can be understood so simply. The male/female binary and largely monocultural perspective used in such studies offer little more than a controlled experiment. Similarly, there are a number of navigational apps designed by women with their lived experiences in mind, especially when it comes to safety and safe routing. But they tend to group the collective experiences of women, neglect the nuances in the lived experiences of women from different cultures and perpetuate the idea that walking is a dangerous activity for women. These apps tend to reflect the views of professional middle-class women from the West in narrowly defined contexts of safe and unsafe spaces based on reported data about street crime, rather than the range of women's navigational knowledge and lived experiences. This is partly a problem of cartography and the need to make selective decisions about what to include on and exclude from maps, but it is also a problem that could be addressed by thinking more broadly about what women from different walks of life want from navigational apps designed for them.

If more women were to set the conventions of what maps are for, there might be more navigational maps based on the lived experiences of women rather than men. But we could do a lot more to understand how gender shapes navigating with maps by thinking more deeply about how gender is perceived and performed in society and across cultures. The way that people might use a map to get from A to B is merely a reflection of this, rather than sitting outside it.

*

The relationship between gender and maps is now an established field of research, and many scholars are studying how gender and culture shape the production and use of maps.[14] Those in this field tread the line between a feminist critique of maps and a feminist approach to mapmaking. From this has emerged an exciting form of counter-mapping that has found new ways to study and represent this relationship, giving it greater depth by linking feminist discussions of gender and maps to those that incorporate the intersectional feminist ideals of equality, including dimensions of race, sexuality and differently abled bodies. This work counters and resists the dominant culture(s) that are commonly represented on the map, and a growing number of people are producing them with the specific aim of highlighting cultural diversity.

Infinite City: A San Francisco Atlas (2010) is a book of maps produced by the writer, historian and activist Rebecca Solnit.[15] It offers insight into how to map the diversity of urban culture that departs from the ways urban life tends to be represented by cartographies of the city. Included are maps representing the city's Black, Hispanic and Indigenous culture, its LGBTQ2IA+ community, and their various and sometimes interconnected fights against displacement and the erasure of cultural practices. These are perspectives central to the history of San Francisco and the land on which it sits, but they are missing from most of the maps that have shaped the common understanding of the city for years: street maps, tourist guides, urban plans, topographical maps and geological surveys. These, like all maps, are not detached from culture, but queer spaces, native toponyms and places central to the fight against gentrification are not represented on maps produced by the likes of Lonely Planet, National Geographic, the United States Geological Survey and Google. This is despite the implicit claim that these maps give an accurate overview of the city.

One of the maps in Solnit's atlas is called *Monarchs and Queens*, made by Mona Caron, Ben Pease and Lia Tjandra. It is designed to

Mona Caron, Ben Pease and Lia Tjandra, *Monarchs and Queens*, map of butterfly habitats and queer public spaces in San Francisco, CA, 2010.

show how queer spaces have shaped the city over the last century. The labels on the map represent the important figures and locations that fostered this community. The butterflies, native to San Francisco, are symbolic of those people who have spread their wings to live a queer life in the city. The underlying cartography of the map may be familiar, but these overlays tell a different story from that embodied in most maps of the city. The writer Aaron Shurin, whose essay accompanies the map, describes it like this:

> This is a map of 'monarchs and queens', of butterflies
> and flutter-bys, or caterpillars in drag and men and
> women with wings. This is a map of transmutations,
> a map of cocoons torn open and places of refuge for

winged creatures, of antennae on the wind and feelers
that feel you up, of mariposas and of Marys. This is a map
of tribes, of flittering things and their gathering spots,
of wings and of wingspread, of extravagant names and
impossible migrations, of will-o'-the-wisps and forces
of will, a map of fritillaries and fairies.[16]

Since this map was published, efforts to place queer cultures
on the map have blossomed further, to visualize them, keep a record
of them and fight for recognition and social change. Queering the
Map, an online community mapping project that started in Montreal
in 2017, is one such effort. It seeks to put queer experiences of the
physical world on the map in any way and any place the contributors want. This way, queer cultures are shown to be what they really
are: diverse and multi-sited, rather than fixed in spaces that are
generally regarded as queer (such as bars, clubs, carnivals and
parades). There are hundreds of thousands of stories of queer experiences on the map, in many languages: stories of love and loss, of
hope and harm.[17] The underlying geographical cartography –
Google Maps – will be familiar to most people, but it is only through
the efforts of the queer community that this particular version suc-
ceeds in its mission to put the lived experience of queer cultures
on the map.

Culture is always represented on the map, so it is important to
question whose culture that is and why some cultures are better
represented than others. For instance, why is it that state-approved
and institutionally recognized cartography reproduces a narrow
view of culture based on a select group of spaces and practices? The
Infinite City atlas and Queering the Map project make it clear that
we could do a lot more to represent and be inclusive of the diversity
of culture. There is much more to culture than the libraries, theatres,
galleries and museums that represent 'culture' on most popular maps
of cities.

Solnit's and Queering's maps highlight the need to make visible the diversity of culture in order to fight to maintain or preserve it. This is a similar aim to that of the Anti-Eviction Mapping Project, a collective of activists fighting displacement and dispossession in San Francisco, Los Angeles and New York by telling the stories of people on maps, through data visualizations and audio and video testimony. In San Francisco, story maps made with local partners are being used to raise awareness of the displacement and dispossession caused by rapid gentrification. They are also being used to build a movement of communities, to create a body of evidence to challenge existing property laws, unlawful evictions and unethical tenancy agreements, and to provide policy recommendations for a city in which there have been immense changes to how land and property are valued.[18]

Counter-mapping like this is a process that subverts the dominant perspectives of cartography, and is part of an ever-growing body of mapmakers seeking to show and make visible alternative ways to see the world and understand its people.[19] This is not just happening in the United States, either. All over the world people are producing counter-maps that emphasize cultural identities in order to fight for social justice, land rights, recognition and much more. And although counter-mapping is nothing new, growing access to the Internet and open-source digital mapping tools that can produce maps to the standards historically set by the professions and institutions of cartography is giving more power than ever to people fighting for change using maps.

As well as those using digital tools to counter the map and put culture on the screen, there are those who favour analogue methods and an arts-based approach for achieving similar aims. Countering the map does not just mean putting different things *on* the map; it can mean doing different things *with* the map.

In March 2021 Sonia Barrett and a group of Black and Brown women from Map-lective produced and displayed their artwork

Dreading the Map in the prestigious Map Room of the Royal Geographical Society in London.[20] This is a counter-mapping project unlike many others. It resists the physical form of cartography by taking selected paper maps of the United Kingdom, the Caribbean, Africa and the United States from different periods, shredding them into strips, plaiting them into dreaded sculptures and hanging them from the ceiling. This methodology, which is grounded in community participation, highlights the interconnections between Black UK-Caribbean cultures and the colonial relations from which they have emerged. The result is a three-dimensional counter-mapping project in which the twists and turns of each sculpted map-dread represent movement and relations between these regions, connected as they are through a colonial past and present. But the work goes further still by countering the norms of the Map Room itself, which is situated in the colonial legacy of the British Empire. The fact that maps of the United Kingdom and the Caribbean were dreaded and displayed by a community of Black

Map-lective, *Dreading the Map*, 2021.

women in a space usually characterized by state-issued maps and the white male figures of empire shows that this was more than just a project to put cultures *on* the map. Instead, it used maps, cultural techniques and community practice to produce a new way to look at maps in this space, to raise questions about whom they should be produced by and how they should be displayed.[21]

Cultures of the map

Maps, like other cultural objects, do not just represent cultures; they also produce cultures and shape them in diverse and interesting ways. Cultures *of* the map are nothing new. We must assume that shared social norms and values – the make-up of culture – about how to produce and use maps have been around for as long as people have made and used maps. Throughout history, these norms and values have had a distinct geography, have been influenced by social demographics and have been affected by wider societal practices. The same remains true today. The teaching of the use of maps in formal education, the way forms and conventions emerge from professional and amateur cartography, and how maps become bound up in specific leisure activities, all constitute cultures of the map.

Since maps became available to the masses in the nineteenth and twentieth centuries, cultures of the map have grown immeasurably and broadened in their definition. These are no longer niche subcultures, either. In some cases, mapping cultures have gone mainstream. Many people will have been made aware of the mainstream's love of maps during the COVID-19 pandemic, with the maps that people hang on their walls becoming a common backdrop for video calls. This is very telling of the ways maps are appreciated and displayed by their owners. People like to show off their maps, it seems.

In the twenty-first century there has been an exponential rise in cultures forming around the map, owing to the popularity of the Internet and the explosion of digital mapping technology. There are now many more mapping cultures than I have space to write about here, so I want to focus on both long-standing cultures that have changed over time alongside changes to mapmaking and use, and newly emerging mapping cultures that highlight the impact digital and mobile maps have had on social and cultural life.

Map-collecting cultures are a good place to start because they represent a diverse group of amateur and professional 'cartophiles' with a long history dating back to the fifteenth century, when maps were collected by royal courts and scholars for the purposes of government, exploration and academe. In the seventeenth century private collecting emerged as a hobby, initially taking off among the wealthy but then opening out to the masses as the cost of producing maps was reduced, first by copperplate pressing and then by lithograph printing. From those who hunt down historical maps at auction and in antiques markets to specialists building a collection of specific maps, and completists whose passion is driven by an aim to collect an entire series of paper maps, there are now many variations of collecting culture.[22]

Individuals, communities and institutions devote time to building and learning more about their collections. They house them in special places, on walls, in dedicated bookshelves, in filing cabinets or in entire rooms. Over time, these collections become part of who these people are; they become a material trace of their life history that tells a story of their interests and expertise. For David Rumsey, the owner of one of the largest private collections of historical maps in the United States, it has become a lifelong passion and – since the digitization of his collection began in 1995 – a way to share this knowledge with the wider world through the Internet. To date there are more than 115,000 maps and related items available to view online in Rumsey's collection.

For such institutions as national libraries, royal societies and universities, maps acquired from collectors have come to represent significant bodies of geographical and historical knowledge. They are bound up in the histories of these places, with rooms, spaces and named collections. The u.s. Library of Congress houses the largest collection of maps in the world, holding approximately 5.5 million catalogued maps. Equally, the British Library is proud of its own map collections, especially the King George iii Topographical and Maritime collection, which was donated by George iv in 1828. It is now housed in the King's Library, an enclosed private space with exterior shelves that is the grand centrepiece of the British Library itself. The library claims that this map collection is 'one of the world's most important historical resources', which is both a bold statement and a reflection on how the institution views itself.

Collecting cultures are more than just material. Collectors talk to one another, sharing knowledge and maps at community events, at public talks, in specialist journals, in private and, of course, on the Internet through forums, blogs, community websites and social media. Membership organizations such as the International Map Collectors' Society (1980–), the Washington Map Society (1979–), the Internationale Coronelli-Gesellschaft für Globenkunde (International Coronelli Society for the Study of Globes; 1952–), the Map Collectors' Circle (1963–75) and Nihon Kotizu Gakkai (the Antique Map Society of Japan; 1995–2007) sprang up across the world in the second half of the twentieth century, offering the chance for collectors and enthusiasts to gather and publish under one roof.[23] Members can enjoy regular meetings, in person and online, and access to specialist journals and archival material. Perhaps more importantly, such societies legitimate map-collecting culture by offering a space for learning and socializing, and thus establish common norms and values relating to what is expected of and by members.[24] These are not closed shops, by any means – most are very inclusive – but boundaries are nonetheless set as to what map-collecting culture is and how to go about it. How to label, catalogue, store and present a map collection 'correctly' are lively topics of debate of which any newcomer quickly becomes aware.

The world of map-collecting is serious business. Maps and entire collections change hands for large sums of money in specialist shops, fairs and auction houses, so expert dealers and certificates of authenticity have become a necessary part of the culture, to prevent fakes and fraudulent practice from polluting the market. One of the most famous and expensive maps ever sold was the Waldseemüller map of the world (1507), which was acquired by the U.S. Library of Congress for $10 million in 2003. This huge wall map, measuring 12.8 × 2.5 metres (42 × 8 ft), made by the cartographer and scholar Martin Waldseemüller and his team, is said to be the first to name and place 'America' after the return of the Italian explorer Amerigo Vespucci's expedition in 1505. It is little wonder it was so highly valued by one of Washington, DC's premier institutions and holders of American history, despite the fact that this map is tied up in a specific colonial history of exploration and knowledge production.

Such sales are not just the preserve of old maps. Modern design classics, such as the various iterations of Harry Beck's iconic diagrammatic map of the London Underground, and map-inspired artworks produced by the likes of Grayson Perry, are highly valued in auction houses alongside other works of art. Indeed, a thriving cartography-inspired sector of the art market has made maps prized commodities for the wealthy as much as for the average buyer looking for a printed replica to hang on the wall.

Connecting these people is a shared love of collecting, displaying and discussing maps (or, in some cases, a love of the money that can be made through trading maps). This is not a culture in the classic sense of one that is national or ethnic, but instead a multi-culture of enthusiasts looking for ways to indulge their interest in maps. Among its various factions, map-collectors share values about maps that distinguish them from others, who may regard maps simply as a practical means of finding their way from A to B.

*

Turning our attention away from the map itself, towards how and where maps are used, we can also see other types of collecting culture that are shaped by the map. These include those that mix a love of maps with that of finding and collecting hidden treasures.

Geocaching is the present-day, digital equivalent of the treasure hunt, with less potential to find life-changing loot and a greater chance of making friends. It is based on the premise that physical items – usually small boxes containing a notebook, trackable tokens or a scroll of paper – can be hidden and assigned coordinates that are represented on a mobile map alongside a clue to their location. Since Dave Ulmer of Beavercreek, Oregon, hid the first geocache on 3 May 2000, geocaching has become a popular activity all over the world. According to Groundspeak, the Seattle-based company seen as the founder and administrator of geocaching, there are now over 3 million geocaches hidden across 191 countries, more than 36,000 community events each year, and more than two hundred geocaching organizations worldwide. This is no niche activity for map nerds. It appeals to those with a completist and competitive mindset as much as it does to puzzle enthusiasts, families looking for a day out, and those who want to engage with their environment in a new way. I've tried it myself in Europe, the United States and across the United Kingdom, and know well the gentle thrill of finding a hidden gem in plain sight of otherwise oblivious passers-by, the physical dexterity needed to reach some of the best-hidden caches, and the difficulty of deciphering some of the clues.

As with other map-collector cultures, geocachers are a part of a community with shared ethics of hiding, recording, exchanging and taking care of one another's hidden treasure. Caches must be stored correctly, protected from the elements and returned exactly the way they were found. Clues must conform to the strict difficulty classification system, and responsibility should be taken to repair or replace any damaged or missing items for future users. And, because every find is recorded and logged on a public ledger, those who fail to abide by the rules can quickly run into trouble with the community.

This is also a vibrant culture of collaboration and exchange, where local, national and international events are organized as a way to meet other geocachers, swap stories of hunting victories and missteps, exchange knowledge and generally socialize with like-minded people. The meet-ups I attended in south London during the summer of 2015 exemplified this community spirit. I was warmly welcomed into the group with discussion of hard-to-find caches, mapping inaccuracies and geocaching etiquette over pints in the pub. Then there are the online forums, websites, podcasts and social-media channels, officially affiliated and not, which provide spaces to talk shop, but also to raise issues faced by the community or make suggestions as to how improve the activity and refocus the values of the group. As with much online discussion, the threads of these forums range from serious discussion about the long-term direction of geocaching to posts written primarily to vent rage at missing or mistreated geocaches, difficult clues, or the social and legal obstacles that must be overcome when finding certain caches.

Geocaching is undoubtedly a popular mapping culture that combines elements of gaming, problem-solving and sense of community, but it pales in comparison with the popularity of the Pokémon Go phenomenon, which at its peak in 2016 saw 232 million users hunting down virtual Pokémon overlaid on the world. This was a cultural phenomenon around the globe, and a mapping culture at that. Central to the interface are the egocentric mobile maps that shift as their users do around parks, streets, shops and playgrounds, to hunt down and collect virtual Pokémon with which to battle opponents. Along with smartphones and the Internet coverage needed to play, maps are fundamental to the way this game works because they show players where to look and who is nearby.[25]

For a time, there was a flurry of reports of people gathering en masse in usually quiet spots, phones in hand, seeking to build their collections. Sometimes entire city centres were said to be overrun with players seeking to 'collect 'em all' as locally organized events went viral after being promoted online. This was the case in San Francisco on 21 July 2016, when a local community event advertised on Facebook quickly escalated into a 9,000-strong swell of people moving through the city, stopping traffic as they went. As well as coincidental gatherings of people, efforts were made to organize officially sanctioned large meetings for like-minded Pokémon collectors. On 2 July 2017 the first Pokémon Go Fest was held in Chicago's Grant Park, attracting an estimated 20,000 people. The excitement was, however, short-lived as network failures led to many people not being able to access and play the game. Some even successfully sued the game's developer, Niantic, for failing to provide a promised service.[26] Since then, the festival has overcome its teething problems and expanded to other countries, developing new formats including Pokémon Go Safari and Pokémon Go Community Days, as elements of the game have changed.

Although the buzz around the game has dissipated from general view, Pokémon Go remains popular among a core community of players who now have many ways to get involved with community events and discussion. These range from officially organized in-person events to local meet-up groups and online forums. There is even an organization called the Silph League, which aims to connect the world's Pokémon Go communities by mapping them. And, as with other large gaming communities, spin-off cultures interested in specific elements and characters of the game have emerged that have given life to new features and gameplay as they are picked up by the game's developer.

Pokémon Go and geocaching are two examples of mapping cultures that have formed around activities that tie maps to collecting culture and outdoor leisure activities. They are not so obviously cultures about the map, but take the map away and they quickly become something else. It is as if the mapping element itself helps to bind these communities together. I want to explore this further in a final example of mapping cultures, where mapping has been the thing that stitches a community together through charity and altruism.

Socially minded communities of all kinds have emerged through a shared use of maps or the making of maps. In the humanitarian sector, maps have become a popular means for people to contribute more than a monetary donation. This is because maps have value for humanitarian organizations and aid workers on the ground looking to understand the impact and extent of disasters, and where aid needs to be directed to.[27] Through 'mapping parties' organized by aid organizations, people are encouraged to come together and produce maps using open-source mapping software. The maps are made by comparing satellite imagery with what exists on freely available online maps, through a process of adding missing details.

I attended mapping parties in London for three years between 2013 and 2016, and observed that these events were humanitarian efforts designed to produce maps as aid, but also social events fuelled by idle chatter, laughter, pizza and beer.[28] Through the work we did together and the conversations we had, it was clear that this was a culture of mapping that attracted like-minded and enthusiastic people who wanted to help by producing maps, but who also wanted to enjoy the process. And, as with the map-collectors, geocachers and Pokémon Go players, these people formed a mapping culture with its own identity and ways of operating. There were said and unsaid rules about how we should go about mapping, key members of the organizing team telling us what to do, and the same faces attending each month. Ultimately a social bond was created through being together and the feeling that we were making a difference. We may question whether the feel-good factor of these events translated into making a real difference on the ground – the jury is still out on this – but what was not in doubt was the community that was created around this form of mapmaking.

*

Doreen Massey describes the problem of putting culture on the map adeptly when she says, 'Loose ends and ongoing stories are real challenges for cartography.'[29] 'Culture' is often used as a place-holder term to simplify and categorize different walks of life, but it defies clear definition. It is a word and a concept that tries to do the impossible and make neat sense of how you live your life compared to how others live theirs. Mapmaking in any form is arguably not equipped to represent all the ways we live our lives, especially at a time when there is so much interaction and evolution between cultures. Perhaps this is why mapmaking moved away from its cultural identity and towards a scientific one. Location pins, place names, gridlines, coordinates, symbols, colours and other cartographic characteristics offer clues as to what cultures are on the map, but they can hardly do justice to the complex social systems that we call cultures. Instead, we are left with partial mappings of culture, leaving map-readers to decide for themselves the significance of what culture is represented on a map.

To understand culture and the map, it might be better to recognize and study cultures of the map. Looking at what people do with maps and what communities form around them provides a wider lens for understanding the connections between. As I said at the start of this chapter, culture never left the map and maps never left culture, despite the strong association between maps and science today. This is something to keep in mind when reading or making a map, or observing someone doing so. These are not activities separate from culture, but rather activities continually shaped and reshaped by them.

Adalbert von Rößler, illustration of the Berlin Conference of 1884–5
with a large map of Africa on the wall, *Allgemeine Illustrierte Zeitung* (1884).

5

Maps that Make the Money Go Round

The Berlin Conference (November 1884–February 1885) kicked off the so-called scramble for Africa, whereby European superpowers vied to secure African territory by offering false protection in exchange for land treaties. It is one of the best-known examples of colonizers using maps to legitimize their land claims. At the height of imperial rule, 'discovered' lands were quite literally divided up using maps commissioned by the state and trading companies with little regard for the people who might be living there. During the conference, which was organized by the German chancellor Otto von Bismarck, European leaders – many of whom had not set foot on the continent – laid claim to African lands by first agreeing that treaties signed by national and regional leaders in Africa were evidence of land ownership, then by drawing up land maps based on these agreements.

In one striking case of mapping for economic gain, conference delegates declared large parts of central Africa (present-day Democratic Republic of Congo) a neutral zone so that trade into and out of the region via the Congo River would be unimpaired by the agreed-on sovereign territories. This effectively became an international zone of trade imposed on the region by outside powers at the expense of the claims on that land of Indigenous people – who, we should not forget, had already suffered a great deal from the effects of the slave trade and colonial rule. Although scholars have since rethought the significance of the conference in the history of imperial rule (large parts of Africa had been colonized before the conference took place, and it is really the symbolism of the event that has stuck in people's minds, rather than the complex reality and its varying impact[1]), there is little doubt that this exercise in land-grabs reinforced the view held by the European colonizers that Africa was a blank slate for their economic development and that maps could help to legitimize the process of laying claim to it.

Demarcating and claiming land with maps continues to legitimize land-grabs today.[2] This is because placing over land an abstract geometric layer that shows clearly what is yours and what is mine remains an effective way of exerting power over people and places. Maps continue to have a powerful economic function in the capitalist economy; this is one reason governments and industry still invest heavily in cartography. Maps lend legitimacy to the prevailing world view that land exists to be commodified in order to support this economic system. But this is by no means the only view of what land is for. We could consider the far wider ways in which humans are tied to the land and what it provides us with – social and spiritual connections, a place to call home and a life-support system, for example – but the crude binary between humans and the Earth has nonetheless become the dichotomous view that shapes capitalist societies around the world. Maps are an especially useful technology for reinforcing this perspective because they offer a view from nowhere, backed up by science, that suggests what they represent is simply fact. If it is on the map, it must be true. Candace Fujikane reminds us that cartographic maps have worked so well for colonial and capitalist expansion for centuries precisely because their so-called scientific characteristics of scale, gridlines, keys and coordinates fix space on the page or the screen and depict clear boundaries and marked differences for the purposes of separating and controlling the use of land.[3]

The conflicts and tension that emerge from the large-scale extraction of natural resources offer clear examples of when carto-capitalist ideology meets the reality of people's lives. The story is very similar the world over, and it goes something like this. Through a period of spatial planning and discovery, surveyors and cartographers contracted directly or indirectly by government or multinational corporations are instructed to find and map natural resources in situations where the possible profits of selling them outweigh the possible costs of extracting them. Extracting resources using this rationale is said to be an essential way to boost economic growth and development. If those resources are found on land that has been claimed by others, either through historical and Indigenous lineage or through legal forms of ownership, government and corporations will work tirelessly to secure the rights to extract from that land. Those who dispute these aims are accused of getting in the way of progress and development.

Claiming these rights is not always a simple or cheap process. It can involve years of conflict, consultation and compromise, due legal process, trade-offs, pay-offs, untold financial resources and, in some cases, underhand tactics and corruption. Maps are important in this process, forming part of the scientific and economic evidence used by government and industry to support the benefits of extraction, as well as increasingly forming the opposing evidence used by those who hold rights to the land to support their claims to remain so. These battles are not won by maps alone, but maps can play a significant role in legitimizing land claims, especially when they are produced by organizations and institutions that carry a mark of authority. This is exactly what has happened in Indonesia.

In 2007 the Indonesian government passed a law (no. 26/2007) that permitted its departments to produce planning documents and maps to show the economic potential of land across the country. This was part of a concerted effort to boost economic development in Indonesia by making the most of its vast natural resources.

In effect this meant the government had the power to determine which land was designated for what economic activity. Despite the provision in the law to protect public and community interests in the process, these acts of designation led to large-scale landgrabs by private investors of forested land to be used for agriculture, most notably for the production of palm oil. When existing landholders disputed the government's claims, their protests were often nullified by the law, which favoured national economic growth. In some cases, the first time local communities heard that their lands were being designated for national economic activity was when the companies moved in to set up their production facilities.[4]

By 2010 it had become clear that this practice was causing spiralling conflict over who had the right to access and work the forests of Indonesia. This was not helped by the fact that two government departments, the Ministry of Forestry and the Ministry of Environment, had contrasting understandings of land use and ownership based on different surveying and mapping techniques, leading to permits being granted with contradictory conditions about what could be done with the land. There are reports that as many as 85 different land-use maps were being used for some 34 provinces, which highlights the scale of the problem.

Much of the conflict unfolded between the islands' Indigenous peoples and corporate agriculture, resulting in localized protests and violent clashes as each attempted to stake their claim to the land. These are two groups that hold very different views on how forested lands should be used, have unequal access to and support from national and regional authorities, and do not have the same knowledge of the bureaucratic process. Since corporate agribusiness was closely tied to the development goals of the state, it could count on government support to protect its land claims. This included turning a blind eye to intimidation tactics and acts of violence, and even supplying government security forces to act on the businesses' behalf. Indigenous communities were clearly disadvantaged when advocating for their own claims, both in terms of legitimizing them in the eyes of the state and ensuring that they could do so safely.

Seeking to address the rising unrest, the president at the time, Susilo Bambang Yudhoyono, responded with the One Map for Indonesia policy, which aimed to produce a single freely accessible digital map and online portal that showed precisely how land had been zoned by spatial planning laws and who had the legal right to access it.[5] When the map was launched in 2018, it was said that it would clearly identify overlapping land claims and clarify who had the rights based on a single consolidated geospatial database. Yudhoyono's successor, Joko Widodo, called it 'One reference, one database, and one geoportal, which essentially is set to prevent any overlap, to give certainty, to give clarity, and to have consistency in building this nation.'

From the perspective of government and industry, the policy was an attractive prospect; finally, land disputes over zoning and licensing arrangements could be resolved using an authoritative map that could support their aims and protect their interests. The map's reception by Indigenous people and local communities was less positive, and many claimed the map would erase the land rights they thought they had, and even make it illegal to practise traditions on lands with deep cultural significance. This reaction was exacerbated by the decision to allow data from participatory mapping to be added to the central database. What was initially hailed as a way to build inclusion into the policy turned into a disaster that created further imbalance of power, when the geospatial data from these maps was rejected because it was deemed inaccurate or did not conform with the data standards of the base map. Kasmita Widodo, who was in charge of the Ancestral Domain Registration Agency, an organization set up to recognize and protect Indigenous land rights, said, 'The Geospatial Information Agency said that our maps are not adequate, not valid according to their standard operational procedures.'[6]

Much Indigenous territory is still not recognized on the One Map for Indonesia, which, we should remember, was designed specifically to clear up land disputes with Indigenous peoples. By wiping them from the map, the government has effectively failed to recognize them as having legitimate rights to land, and dismissed the idea that they could contribute to the national economy.[7] Maps have the power to convince the viewer of the reality they depict, which makes them vitally important in land disputes. In the Indonesian case, maps commissioned by government departments carry weight in disputes over land, being certified by the geological and geographical sciences and increasingly produced using innovations in digital mapping, which claim to objectively identify areas of forest that are suitable for agriculture.[8] They are also accompanied by documentation that supports these claims and argues that they are made in the name of national economic development. The maps produced by Indigenous peoples, which tell a very different story about these places, struggle to compete in a world that favours cartographic rigour over their ways of seeing the world.

Maps like One Map for Indonesia create a clear imbalance of power between the interests of the state and private industry, and those of local communities, who are less likely to have professional, institutional and technological backing. Today, states do a lot to show outward support for community land claims, but ultimately they have the power to ignore or appease these claims if they obstruct the goals of the state, which are increasingly tied to the goals of capitalism. We see this time and again where Indigenous and other communities experience long historical land claims being quashed by states using maps to secure land for the extraction of resources.

But it is not always so simple, and these once-established power imbalances are beginning to become undone as local communities, aided by activists and NGOs, look to counter-mapping techniques and digital mapping technology to fight for and win their claims to land.[9] In Indonesia, the potential of using aerial drones and digital mapping platforms to win these fights has recently been recognized. Between 2011 and 2015 the scholar and activist Irendra Radjawali and his colleagues were successful in training local communities to use drones and make maps to settle land disputes over bauxite mining in the West Kalimantan region of the country.[10] High-resolution photographs taken using drones were stitched together and mapped with software to show where mining companies had built, polluted and caused environmental damage on lands outside their licence agreements. This was enough to convince local authorities, and in several cases mining operations were shut down as a result.

There is a history of participatory counter-mapping in Indonesia that dates back to the 1990s, so these methods are not completely new.[11] There remains some debate over who in these communities has access to the technology and the knowledge to use it, and how cartographic mapping perpetuates the capitalist world view of human/Earth relations that is at odds with Indigenous views.[12] Nevertheless, the West Kalimantan case highlights how relatively inexpensive drone mapping technology can lend legitimacy to land claims, which is often the primary concern of communities fighting for recognition and to prevent further ecological and cultural damage to the place they call home. High-quality counter-maps will not alone win the battle with state-led economic development, but in this case, at least, they have proved able to equal the power of the stakeholders.

Using maps to designate zones for economic activity is an established practice across the world, but the specifics of where and how it is done, along with the reaction to it, matter to the way we tell the story of maps that shape the economy. The examples laid out here cannot be used as definitive evidence that maps shape economic development in the same way everywhere. However, they are further illustrations that maps are what we make of them and do not offer universal views of the world. Colonial powers using maps to shape economic interests in Africa, the Indonesian government using maps to shape economic development in Southeast Asia, and Indigenous and local communities using counter-maps to shape a fight against this, are all further examples of how and why we create maps with different aims and intentions in mind. Those aims are in some cases to exert the power of national economic policy, and in others to subvert that power and highlight a different understanding of the world.

*

Cartography has been an invaluable instrument for economic policy throughout history, but the relationship between maps and the economy is far more varied than that. This is especially true today, when maps feature in so many economic transactions, ranging widely from the acquisition of land and resources to the purchasing of property and the operation of transportation services, and use in the digital and gig economy. Prospecting for resources, buying a house, hailing a taxi, running a transport network, tracking online orders, locating bike-sharing services and finding a business on the high street are all economic activities routinely shaped by maps. Among other things, they tell us where things are and when they will reach us, they dictate ownership and they are used to estimate transaction costs. There is a history of mapping in these sectors, but in recent years these uses have been accelerated by the rise in mobile maps and what we might call the egocentric cartographic imagination bought about by GPS technology. Many people have become accustomed to maps that provide a lens through which to visualize aspects of their spending.

Perhaps most notable has been the rise in the economic value of location-based data that is viewed on maps. This is the data that many of us now routinely give away every time we use our digital devices. Location data is big business, in most cases commanding far more than the maps themselves. When we consider all the apps for which maps and location data are key, and the fact that mobile devices rely on location-based technology to operate, it is easy to see that this amounts to a market worth billions of dollars each year.

Maps have always been a product of data. The process of collecting information about the world and representing it on a map is as old as the map itself, whether it be the systematic collection of data through census or scientific study, or the plotting of information about observed people, places and objects. Data is still used to create maps, routinely collected, sorted, cleaned and aggregated by humans and machines to produce maps that place it within precise geometry on the screen or page. The wealth of data that is now available has also created the conditions for maps that are constantly being updated, in some cases many times a day, or even each time they are used. In 2020 the vice-president of Google Maps, Dane Glasgow, announced that it made 50 million updates to its maps every day owing to changes in the data set; this is astonishing considering how infrequently maps have been updated in the past, but also not surprising in today's digital world, where constant iteration is built into the design of technological innovations. And because maps are now subject to data changes at any time, it is true that whenever we load a digital map, we may not see exactly the same map we saw the last time we used it. What we see is just the latest version in an ever-expanding spatial database. This raises many questions about the authorship, authenticity and reliability of maps, questions that may be asked in instances of ordinary or extraordinary use of maps, where claims are made based on what is on the map. This has implications for an economy that relies on an established order of things, and increasingly on widely used digital maps that represent that order.

Mapping giants play a fundamental role in the global economy because being represented by them acts as a shop window to the world. Unsurprisingly, the world's largest shop window is Google Maps, with around 80 per cent of the market and more than a billion users. Those making up the remaining share – Apple Maps,

Here WeGo, Baidu Maps, OpenStreetMap and others – are gaining ground but pale into insignificance in the everyday consumer market. To be included on Google Maps is to be visible to the largest market. To be left off it puts a business at a serious disadvantage in today's economy, with market analysts citing Google Maps as one of the primary ways that new customers learn about a business. When map searches are linked to web searches, this matters even more, because the more a business is searched for on Google Maps, the more likely it is to appear near the top of a Google search results page. The practice of search-engine optimization – getting your web page seen among the crowd in online search results – has spawned an industry of both experts and scammers claiming to make a business visible on maps and in search results alike.

When Google makes changes to its database or terms of use, business owners can find themselves wondering why they have been taken off the map overnight, and how this might affect their bottom line while they find ways to get back on. An area of digital marketing strategy is devoted to troubleshooting these changes, made up of online forums, professional video tutorials, industry articles and expensive consultancy for businesses looking to get back on the map. Conversely, there are those who are employed to help a business get off the map, as a result of going bust, moving premises or providing services that it does not want to be made public. In 2016 an abolitionist sex-work group made the names and addresses of dozens of sex workers in Germany, Belgium and Denmark publicly available on Google Maps in an attempt to name and shame the industry that it saw as perpetuating the trafficking of women in Europe. This was met with scorn by some working in the industry, who sought to have the information removed from the map as quickly as possible. The Professional Association of Erotic and Sexual Services in Germany called it a dangerous act of violence that put the lives and livelihoods of workers at risk, and did not take into account the wide range of reasons that sex work is undertaken. This case shows that maps that shape the economy have the potential to be about far more than just money.

Then there is the shady practice of manipulating the spatial database in an attempt to remove rivals from the map, or changing key pieces of information in order to put off potential customers. This is what happened in Buffalo, New York, in 2010, when the Barbara Oliver & Co. jewellery store had its Google Map listing tampered with by a competitor, who edited it to show that the shop was 'permanently closed' and swamped the reviews with negative comments. Google has since put a stop to this trend by making it more difficult to edit another business's public information, but this has not put a stop to the malicious editing of the map. Scammers and rivals are having to find ever more ingenious ways to make their alterations, many of which rely on manipulating the keyword database that supports Google's algorithmic ranking of search results. By flooding the database with fake businesses in the same area and tagging those businesses with as many popular keyword indicators as possible, tricksters and fraudsters alike are able push out legitimate businesses from search results shown on the map. This is an example of how people are gaming an economic system dubbed 'linguistic capitalism', whereby the value of keywords used for ads is the overwhelming factor in determining the order of search results.[13]

Mapmakers have long looked to the public to inform them of their errors. Web mapping platforms are no different, except that today, rectifying these mistakes has become the responsibility of the contributor. This follows many other digital services where the 'prosumer' model (in which the user is both producer *and* consumer) has become the favoured business strategy, but it does provoke questions over where responsibility for errors should lie. As it stands, large web-mapping platforms take no responsibility (legal, ethical, moral or otherwise) for ensuring that errors on the map are put right, even when they benefit enormously in financial terms from having people use their service. We hear publicly about changes to the map only when the maker is trying to show support for a particular cause or to avoid being locked out of a particular market. In some cases, this is a real problem for the average user, such as when old or incorrect addresses buried deep in the database have meant people's homes becoming de facto business addresses that attract unwanted trade to their front doors, or when businesses are taken off the map without their owners being notified, leaving them unaware of the revenue they may have lost.

*

Aside from being used to create and update maps for commercial interests, data has also become an economic product *of* maps and how they are used. This marks a shift in the relationship between data and maps. No longer are maps just the representation of data; now they are a producer of data as well. This is true of map-only services, such as in-car navigation systems, as well as for digital services in which maps are embedded as a key interface alongside other features, and where geographic information plays a central role in how they are used, for example in dating apps, ride-hailing services and real-estate platforms. In these latter cases, location data is enriched significantly by how it relates to other user data. Knowing how a map is being used, relating this to where someone is using it, and linking this to their name, profile, financial details, the time of day and how they have used other features, provides the operators of these services with a wealth of valuable information about their customers. As with most digital products and services, the aim of collecting (location) data is to find monetary value in it, usually through selling it directly as a data set, or indirectly as packaged analysis offering insights and market advantages.[14] In recent years the industry's appetite for data collection has grown exponentially as it is used to feed machine-learning systems designed to find patterns in large, disparate data sets, and ultimately to predict and shape possible future behaviour.[15]

In 2006 the British mathematician Clive Humby called data 'the new oil', by which he meant that data is fast becoming a foundational commodity for the global economy.[16] In 2020 Lord True (Nicholas True) began the British government's Geospatial Strategy report by saying that 'location data is the coal and iron fuelling a new revolution.'[17] In a world where more and more activities are carried out on digital devices connected to the Internet, it is easy to see how these statements might have some veracity. Every time we engage with digital and networked technology, we generate information about how we use it. That information can be aggregated and analysed along with the information from all other users to paint a general or very specific view of how and why the technology is being used. Some of the information we generate is obvious. It includes behavioural data: what we did with the technology, what features we used and how we used them. Some is less obvious: information about how long we used the technology for, what device we are using and where we used it, for example. Then there are the hundreds of other data points that are collected but that we rarely consider. This is the wealth of information that is collected by tracking users' behaviour across and in relation to other technology, say between apps or websites through the use of cookies. Taken together, the technology we use can build a sophisticated picture, not only of how we use it, but of who we are. This is a very valuable asset for technology companies because it can be used to improve their services, but can also be sold on for a profit to those looking to target specific consumer markets. In today's digital economy, that primarily means advertisers.

There are early signs that we *may* have reached 'peak' data collection, that the days of tech companies mining our location for profits without consequence are coming to an end. We see these signs in the public debates about data privacy, in the tentative legislation being brought by governments around the world, and in how the business models of technology companies are adapting to the shift towards the privacy-conscious consumer. Nevertheless, the appetite for data will not disappear any time soon, even if the modes of collection are beginning to shift. The economy relies on it too much for it simply to stop.

Location-based data has become central to the economy since the early 2010s as more and more people have come to see the value in knowing where someone or something is. By some estimates the market for buying, selling and analysing location data was worth $14 billion in 2021, and is expected to increase by 15 per cent by 2030.[18] Retail businesses, insurance companies, banks, universities, governments, planners, NGOs and more are all interested in harnessing the power of location for a range of reasons. If you know where someone is and where they have been, you can tell a lot about what kind of person or consumer they are and what they might like to buy, do or vote for next.

Correlation is king in this world of data analysis, and it is significant patterns of behaviour derived from the analysis of location and other mobile data that demand the biggest asking price. Take, for instance, a sports team looking to negotiate its next sponsorship deal. The club can place a high value on the location data it holds about its fans, collected through the club's app, which might collect its users' journey data to and from the stadium. This could give sponsors an insight into where supporters live, as well as where they go before and after games, and that is highly valuable to any company looking for market segments to target with advertisements for the latest product or service.

In election campaigning, this form of micro-targeting was rife in the run-up to the U.S. presidential election in 2020, with both sides using targeted ads on social media based on their supporters' most frequented places. In one famous case, Steve Bannon, a former strategist for the Trump campaign, was found to be running a localized campaign with the political advocacy group CatholicVote, which targeted Catholic churchgoers in the states of Iowa and

Wisconsin.[19] Using location data obtained from telecoms carriers, the campaign team was able to identify who among the local electorate had been to church in recent days and weeks, then target their social-media feeds with ads for the Republican candidate, the incumbent president, Donald Trump.

These forms of location-based marketing and micro-targeting are not new. The harvesting of mobile data dates back at least to the year 2000, and the use of location for advertising to the birth of marketing itself. Yet claims to offer 'precision marketing', 'advanced solutions' and 'richer insights' gleaned from patterns in location data still abound to the point of hyperbole in the digital analytics industry, and promises to reveal key insights using ever more sophisticated analytics technology are rife. Central to this market are data brokers, who hold and facilitate data exchanges, and data analytics companies, which claim to be able to offer bespoke analysis of this data and how it can help you.[20]

We can expect these assertions to continue as long as people see value in this data, but we should be cautious of claims that patterns found in data can reveal everything about human lives. Although the 'datafication' of everyday life grows every year, there is no end point where finally enough data will have been collected to tell us everything we need to know.[21] Besides the fact that the survival of the industry requires the perennial collection of data and innovations in analytics, the insights of even the most sophisticated data analysis are regularly proven wrong, misleading or incomplete. Rarely, for instance, do patterns of travel behaviour reveal why we went somewhere, which route we took or how long we spent there. Identifying patterns of footfall or a regular journey to the stadium for a season-ticket holder, for example, reveals almost nothing about *why* fans support a team or about their lived experiences in the build-up to and during the game.

Qualitative researchers have long shown that quantifying human behaviour and identifying patterns in order to find meaning gets us only so far. The contexts, nuances, idiosyncrasies and complex relations of human behaviour demand different modes of analysis and an acceptance that some slices of life will always remain hidden from view, no matter how detailed the data set is. But this is hardly the point; the power of data and the analytics industry stretches beyond the desire to understand human behaviour. As the geographer Rob Kitchin makes clear, data has become an ideology for our age, supported in droves by the most powerful in society.[22] This can be supplemented with the prominence of 'algorithmic reason', which, the media scholars Claudia Aradau and Tobias Blanke tell us, is when society trusts algorithmic systems to be the sole keeper and handler of our data, despite their opacity.[23] What is really at stake here is how these prevailing views are used to shape public life, not only to target potential consumers, but to track, control and discipline people. When it comes to the collection and analysis of location data specifically, it is how and why it is used and shown on maps that matters, because this is how it is seen, understood and shared in ways that people understand.

Aside from the economic advantages of collecting location data, many others have recognized the potential of such data for creating new forms of surveillance, and its ability to invade privacy in new and disturbing ways. We have already seen that maps are being used as interfaces for tracking the movement of people. This is now a common practice all over the world, taken up by states and businesses alike in the name of national security and fair market advantage, and in the domestic sphere by parents using location-based apps to track their children. Fed by anxiety and fuelling household disputes, location data in these cases represents much more than just a point on a map. It represents a complex relationship at the intersection of trust and privacy between parent and child.

Much of the time, the practices of collecting and using our location data go unnoticed as we click 'agree' to the terms and conditions of our favourite services – which do tell us how our data *may* be used, albeit using legal jargon. However, every now and again we are exposed to the wider impact of how our location is being used by others.

Sharing location data with and selling it to third parties is standard practice for most technology companies. The trade-off, whereby the user accesses a free service in exchange for their data, is something we have all become very familiar with. Nevertheless, in 2020 researchers at the Norwegian consumer council, Forbruker-rådet, published a report detailing how location data from the popular gay, bi, trans and queer dating app Grindr was being shared with third parties other than those named in its terms-of-use policy.[24] This led to an investigation by the Norwegian Data Protection Authority into a breach of the EU's GDPR (data-protection) rules, and the company was subsequently fined €7.1 million for illegally sharing personal data. Grindr had form in this arena, having been found to have shared the HIV status of users with third-party analytics companies in 2018, despite claiming in its terms of use that this was prohibited; it was also found to be easily hackable, so that anyone with a little technical knowledge could locate users on a map based on a distance-measuring process called 'trilateration'.

This should have been a warning to the company, but in 2022 the *Wall Street Journal* published an investigation into Grindr's data-sharing practices, finding that the company had been selling bulk location data from its users to the highest bidder since at least 2017.[25] This raised understandable concerns for users who wanted to keep their whereabouts private, something that is all the more important for marginalized users in some parts of the world, where non-heterosexuality remains illegal or socially unacceptable.

Jeffrey Burrill, a priest and the general secretary of the United States Conference of Catholic Bishops, paid a heavy price when *The Pillar*, a prominent Catholic news outlet, ran a story in July 2021 claiming he had had a series of homosexual encounters, based on location data taken from his phone, obtained via Grindr on the open market.[26] Burrill had no choice but to resign. Beliefs aside, this case shows how personally damaging the sale of data from mapping technology can be. Rather than being a story about how maps shape the digital economy, it quickly becomes a story about how the availability of location data can send shockwaves through a religious group and turn one man's life upside down. It is cases like these that add to the calls for more regulation of the collection, sharing and selling of location data, which as it stands is notably absent, outside the landmark EU GDPR ruling that came into effect in 2018. Even so, all companies need to do to circumvent this law is to ask for consent to share this data in exchange for using a service, which the majority of people are happy to do so that they can continue using it.

The Grindr case and others have encouraged security agencies and civil-liberties campaigners to study the potential impact of personal location data being readily available for purchase. Although such data is shared and sold under the premise that individual users cannot be personally identified, in reality it is possible to tell a lot about the details of who someone is by examining where they go, regardless of whether the data is connected with any personal information (names, phone numbers and email addresses). Grindr and others have disputed this, but, as a series of investigations by the *New York Times* reveals, it is relatively easy to do. If you have access to a database of individual locations and times – two common features of data sets that are shared and sold – it is possible to identify people based on their patterns of behaviour.[27] Location data showing movement between two locations five days a week, and data showing that these movements happen at similar times each day, can be indicative of a home, a workplace and a commuting route. From there it can be relatively simple to decipher exactly who the person is by matching that data with public records. Much of the time these patterns are mapped for the purposes of targeted advertising, but the fact that this data is available to buy has been used by some to claim that it could be used in nefarious ways, for instance by stalkers, criminals and law-enforcement agencies.

Police forces around the world are increasingly turning to location data to fight, prevent and predict crime. 'Location intelligence', as it is dubbed, is both an effective and a contentious practice, depending on who is asked. Advocates claim it helps to map the movement patterns of known criminals and whom they interact with, and that it can be used to predict the hotspots of crime on maps in real time. Critics claim it is an invasion of civil liberties that reproduces historical patterns of discrimination because it involves using the recorded locations of past crimes in order to predict future crimes and concentrate resources.[28]

How law-enforcement agencies collect and use such data is a complicated topic that varies widely, depending on national data-privacy laws and how they are interpreted by local police departments and individual officers. In democratic countries, location data cannot simply be taken from a person's phone just because they have been accused of committing a crime or are suspected of being involved in criminal activity. Investigating officers require a legal warrant to search a phone for data, just as they do to search someone's house or look through their financial records. Even when a warrant has been granted, it is still up to the tech companies to disclose this data or not. Apple has had numerous brushes with the law in the United States when it has refused to grant law-enforcement agencies permission to access personal phone data. The company argues that to do so would create a back door to state surveillance and damage civil liberties.

Nonetheless, there is much evidence to suggest that the police and other law-enforcement agencies have found ways to access and make use of personal location data that bypass the courts. By some accounts the practice is rife. A Privacy International report from 2018 into the 'digital stop and search' practices of the police in the United Kingdom found that 26 out of 47 forces used mobile data extraction technology, including the extraction of location data.[29] Such technology is found in the data analytics market, often demanding a high price and frequently tailored for use by police forces. It includes analytic tools and maps to make sense of extracted data, and physical hardware that is placed in strategic positions to intercept data from mobile phones that come into close proximity with it.

In the United States, stories abound about how state and federal government departments routinely enter into partnerships with data brokers and analysts to buy location data. In 2021 the Center for Democracy and Technology found that the Federal Bureau of Investigation, the Drug Enforcement Administration, the Internal Revenue Service, the Central Intelligence Agency, Immigration and Customs Enforcement, and Customs and Border Protection all had contracts for location data with brokers including Venntel and Thomson Reuters. In some cases, law enforcement has found ways to collect location data directly from mobile phones, avoiding the market altogether. Chicago's police department, the second largest in the United States, has been using International Mobile Subscriber Identity (IMSI) as cellular network infrastructure to intercept mobile-phone data since 2005, if not before.[30] It was also one of the forerunners in the United States to use the predictive policing giant PredPol (now Geolitica), a popular analytics platform that uses location data to predict crime hotspots in real time and represent them on a map that can be used by officers on the ground and operators in the control room.

Law enforcement in the United States has found much value in using location data, and therefore ultimately the digital maps that are used to represent it. The U.S. Department of Justice found that a third of large police departments in the country were using predictive policing technology by 2017, and it is easy to understand why. The technology is sold as a new way of seeing where crime happens and where it might happen next, ensuring a more targeted approach to policing. The marketing from PredPol reads:

PredPol has a precise definition of predictive policing. For us and our customers, it is the practice of identifying the times and locations where specific crimes are most likely to occur, then patrolling those areas to prevent those crimes from occurring. Put simply, our mission is to help law enforcement keep communities safer by reducing victimization.[31]

Equally, the data analytics industry has found law enforcement to be a valuable and repeat customer, willing to purchase its data services and buy into the claims that location data can be an effective tool to fight crime. Annual subscription costs alone can run into the hundreds of thousands of dollars, and there are additional costs for consultancy, analytics expertise and further data acquisition.

This is not, however, a harmonious arrangement that everyone agrees with. Even within police forces, the move towards predictive policing is contentious, and patrol officers are concerned that their relative autonomy and judgement are being eroded by the computational systems favoured by teams of crime analysts working in offices, away from the action. The criminologist Simon Egbert and national security expert Matthias Leese found that risk maps produced by predictive policing software were a key bone of contention, often used by analysts to pull officers away from the areas they had years of experience policing. However, they also found that over time, risk maps were one of the primary means of building working relationships between analytics staff and patrol officers in forces where the technology was adopted, because the two groups shared an understanding of the value of maps.[32] This is a further example of how context matters, and how it evolves. Office analysts and patrol officers looking at the same map can at first interpret very differently what it shows and what it could and should be used for in policing, but over time these differences can be overcome as they learn how to use the map together.

Adding a further dimension, many critics argue that predictive policing technology perpetuates dangerous, racist and discriminatory policing practices. Research shows that such technology is especially susceptible to this because it relies on local crime-reporting data sets to make its calculations, which are skewed in the favour of reported street and domestic crimes because they are more visible than other crimes. Numerous studies have found these data sets to include racial bias because more street and domestic crimes from non-white neighbourhoods are likely to be reported, compared with other crimes, such as fraud, embezzlement and counterfeiting, that are more likely to occur in white neighbourhoods.[33] The result is that more crimes are predicted to occur in non-white neighbourhoods, which creates a self-fulfilling prophecy whereby policing is stepped up but little is done to address the underlying social, political and economic reasons for why these areas have historically experienced more crime than others. Meanwhile, efforts to address crime in white neighbourhoods are under-resourced, meaning that less is reported, which leads the systems to continue to calculate that crime does not happen in these areas. And so, history repeats itself.

Ruha Benjamin, an expert in African American studies, calls for a rethink of predictive policing technology in light of the historical racism in policing.[34] When the premise and functionality of such technology are founded on data sets that are known to include problematic racial bias, this will only ever reproduce historical patterns of institutional racism in the present. Benjamin notes that when we rely on analytics algorithms that cannot be easily explained to police our streets, what we are really doing is reproducing inequality while at the same time abdicating responsibility for it on to machinic ways of thinking, which do not centre ethics or reflection in their calculations.

Algorithmic systems designed to make predictions and decisions using large data sets are increasingly common. The political geographer Louise Amoore argues that such outcomes as racially biased maps of predicted crime hotspots are not the fault of the machines that made them; those machines are after all doing *exactly* what they are programmed to do. The problem is rather that we have become convinced that algorithmically calculated outcomes are, objectively, the correct ones. For Amoore, such systems demand a conversation not only about the workings of this technology, but (and more importantly) about the ethics of their use and the political motivations of those who use them.[35] As accusations of institutional racism continue to be made towards law enforcement around the world, it is high time to give more careful consideration to how location data and the maps that represent it are used to shape approaches to crime. These cases should also raise questions about how maps, location data and analytics are shaping the economics of policing as never before. As with consumer markets, the police and ultimately the taxpayer are having to pay enormous sums for access to this mapping technology.

*

Maps are valued for their capacity to shape the economy in different contexts and at all scales. This can bring benefits to family businesses trying to grow as much as they can to states looking to develop national economies and corporations wishing to monopolize markets. Nevertheless, there are costs – financial, human and environmental – involved in using maps to shape the economy, and these are usually inflicted on the most disadvantaged in society, or on those who do not want to participate in a capitalist economy. We should remember that the maps and data that shape our economy are not just outlines and numbers. They are an important cog in the capitalist machine, and they shape how geographic information is understood and valued by those who operate within it, and how it can be used to extract profits and exert power.

6

Mapping Presents and Futures

In the opening scenes of *Methuselah's Children*, a science-fiction novel by Robert A. Heinlein, first published in 1941, we find one of the earliest fictional accounts of self-driving cars:

> Mary had no intention of letting anyone know where she was going. Outside her friend's apartment she dropped down a bounce tube to the basement, claimed her car from the robopark, guided it up the ramp and set the controls for North Shore. The car waited for a break in the traffic, then dived into the high-speed stream and hurried north. Mary settled back for a nap.
>
> When its setting was about to run out, the car beeped for instructions; Mary woke up and glanced out. Lake Michigan was a darker band of darkness on her right. She signalled traffic control to let her enter the local traffic lane; it sorted out her car and placed her there, then let her resume manual control. She fumbled in the glove compartment.

The novel points towards a future that car and technology companies would have us believe is almost here. Settling back in your seat for a nap is illegal while driving an autonomous vehicle, and our roads remain a messy mix of driver-controlled and driver-assisted cars, but we are already at the point where a driver can switch between manual or automatic modes, and cars are already smart enough able to judge breaks in the traffic. This has been the case on commercially produced cars since 2008–10 and continues apace today, as investors pump money into a growing industry that aims to change one of the most ingrained cultural habits. By some estimates, the global market for self-driving cars will be worth \$2.3 trillion by 2026.[1]

Curiously, maps are not mentioned in the story of Mary's journey. It is as if Heinlein predicted what cars would do for us, but not how much maps would become part of the driving experience in the years after he wrote the book, first through the proliferation of the paper road atlas and then through the mass adoption of the satnav. In today's self-driving cars, maps are at the forefront of the driving experience. They are displayed on large touchscreen consoles and give drivers directions, traffic updates and arrival time, much like the navigation devices that many people have been using for some time. But maps are also integral to the underlying technology that guides these vehicles and ensures the safety of the passengers. It is these developments that provide a window into how maps will shape the world in the coming decades. These maps raise new questions about how and why maps are made, who the map-reader is, and what maps are used for.

If maps are what we make of them, what are maps if it is not us who are reading them? This is a quandary for those thinking about the practices and ethics of machine-readable maps, such as the kind that support self-driving cars and other autonomous vehicles. Such maps are not designed to be read by humans, or even necessarily to be seen by humans. They are based on machine-learning principles, commonly known as artificial intelligence, and they are designed to help computers make decisions under the bonnet, such as when a vehicle should turn left or right, change lane, slow down or speed up, or simply when a robotic mower needs to turn around after reaching the end of the lawn.[2] How, where and why autonomous vehicles use maps still matter, but now we are left to rely on our interpretation of a computer's understanding of the map if we are to make sense of it all. This exemplifies a wider reliance on algorithmic systems and the ways they have come to make important decisions without us really knowing how.[3]

Depending on the manufacturer, self-driving cars use a combination of sensor technology, huge spatial databases and highly accurate digital maps to understand the road and the wider environment.[4] Computers align sensor and image data captured in real time with crowdsourced base maps stored in the cloud, to understand where they are in a process known as 'localization'. The car, in a sense, becomes the surveyor, the cartographer and the navigator all at once, taking measurements, drafting a map, then using it as it goes along. What it produces is not a map that we would recognize or could use in any meaningful way for our own navigation. Some manufacturers call it 'map-less driving' or 'driver-less mapping'. These maps appear map-like, but they differ from our usual understanding. They are three-dimensional mappings of space and time, largely free of the cartographic conventions that we associate with maps. When we look at a rendering of such maps, we see an ethereal world painted in luminous colours, where objects are identifiable but seem to exist in a different reality. Most accurately, this is a metaphorical mapping process that helps us to conceptualize what the car is doing as it learns about the world around it. If any maps are produced from it, say to visualize the process for education or marketing purposes, they are merely images for us to interpret, not for the computer, which understands the world in a fundamentally different way.

The maps that a self-driving car makes can be distinguished from the maps we might use for navigation. They are made with different aims and intentions in mind, and are not tied to social and cultural life in the same way. The way a machine reads them in relation to the world is also different from the way we do. We may well believe in the notion that the world is accurately represented on our satnav, but we are equally capable of holding contradictory

Digital rendering of Lidar mapping technology used in self-driving cars, 2023.

views at the same time. We understand that there are differences between what is shown on the map and what we can see in front of us, from the driver's seat. In contrast, self-driving cars always understand the map to be the territory. For the machine, nothing exists beyond what it has sensed or what spatial information it has been programmed with.

The mobile-mapping researcher Sam Hind describes the process of automated vehicles making sense of the world as one of feeling it out through sensors and cameras, capturing data as it does so, then calculating meaning from it in order to come to a decision about whether an action should be taken.[5] What the car 'sees' or, more accurately, 'senses' is a world with defined geometric boundaries, in which all sensed objects are classified. Unlike human drivers, these machines are always paying attention, which in theory makes them safer. However, autonomous vehicles do still crash, generally because the computation that would trigger any collision-avoiding manoeuvres has taken too long.

The differences between human and machinic navigation come to the fore in the way self-driving cars react to the road ahead. Any obstruction to the route calculated for a journey, in the form of a swerving car, stray cat or person crossing the road, is picked up by a suite of sensors and cameras, and classified. The system then calculates the probability that the classified object will impinge on the vehicle's trajectory. All this is done in a matter of milliseconds, and each calculation is updated as the car moves closer to the object. If the probability that the object will impede the car reaches a calculated threshold for danger, say because it recognizes quick-braking vehicles or a swerving cyclist, the computer will instruct the vehicle to act, whether it be to slow down, stop or manoeuvre around the obstruction. If the probability does not reach the threshold for danger, say because it is classified as an atmospheric obstruction, such as rain or fog, the computer will instruct the vehicle to carry on as if the obstruction were not a danger to the vehicle.

Elaine Herzberg was the first person to be killed by a self-driving car not reacting to her presence, as she walked across the road with her bicycle and shopping bags on 18 March 2018 in Tempe, Arizona. At exactly 9.58 p.m., a Volvo suv powered by technology developed by Uber's self-driving division struck Herzberg. Just 0.2 seconds before, Rafaela Vasquez, the supervisor-driver employed to monitor the vehicle, was alerted and asked to override the system manually – far too late for a collision to be averted. In the criminal investigation that followed a hasty financial settlement to Herzberg's family, Uber's legal team blamed Vasquez for watching a talent show on her phone and not intervening sooner. Vasquez argues that she was merely listening to the talent show (*The Voice*), and waiting for the system to issue the alert, as she had been instructed to do.

Delving into the data that was released to the public, Hind found that the computer system designed to prevent accidents was busy calculating what Herzberg and her bike were, right up to the moment when it was too late.[6] Between 5.6 and 0.2 seconds before the collision, Herzberg and her bike were classified and reclassified ten times as the machine tried to identify them. The system could not make up its mind whether they were a 'vehicle', a 'bike' or 'other' – an indecision that cost Herzberg her life.

What this tragic case tells us is that there are limits to real-time mapping when vehicles are moving at speed. The latest classification, even if correct, does not matter if it comes too late for the vehicle to act or alert a supervising driver. The incident represents a marker for the industry, when the ethical and political questions of automated decision-making came to the fore for self-driving vehicles. And, although Uber has since closed down its self-driving division and regulations have changed in the United States as a result, five years after the incident the legal case has still not been closed. Vasquez maintains that she was the scapegoat for errors made by Uber's technology, and she may still be acquitted of the negligent homicide charges she faces.

Theoretically, self-driving systems work, but we are a long way from the wide adoption of self-driving vehicles. This is not least because of the costs involved, the technical hurdles for making the system work for all road types and all driving conditions, and the necessary changes to driving infrastructure, but it is also because society does not seem fully ready to commit to vehicles that take away the driver's autonomy, nor committed to unravelling the complex legal and ethical concerns they raise. Nevertheless, there is a great deal of hype in this area that, if realized, means we are set to place a great deal of trust in computers to read maps very differently from the way we do. This is not to say that computers read maps incorrectly, nor that there are not great benefits to this different reading – if we were to adopt self-driving cars en masse, it is quite likely that they would significantly reduce the number of road accidents caused by human error – but it does mark a shift in how maps are used to shape and make sense of the world. It is a shift that asks new questions about the politics and ethics of automated decision-making on the roads. The cases where self-driving cars have crashed and even killed people offer a challenge to lawmakers and technologists, as well as to all drivers and passengers, about who must take responsibility. If the car is simply reading the map, and the map is taken as reality, who are we to blame? The computer is, after all, carrying out its task in exactly the way it was programmed to do. As a society, we are keen on technological developments but less so on assigning blame to them when they go wrong, when we are more inclined to find a human in the loop to take responsibility.

As Louise Amoore has argued, machine-learning technology such as the kind that exists in self-driving cars requires a rethinking of responsibility, not as something that can always be assigned to an individual, one machine or a piece of software, but as something distributed among varying social actors and to be reckoned with in our social and political imaginations about what this technology can and should do.[7] She calls for new modes of thinking that go beyond making the workings of technology transparent – a common trope in calls to consider the ethics of machine learning – and asks that we think more carefully about how and why algorithmic decision-making is fast becoming mainstream. As the case of self-driving cars suggests, these are discussions to be had with maps in mind, and while recognizing the different ways that computers and machines might read maps and make decisions.

*

Away from the road, similar 'geo-visualization' technology is being used by those looking to uncover new perspectives on the world. This includes archaeologists using Lidar scanning and machine-learning technology to make new discoveries and to protect finds; urban planners building city-wide three-dimensional maps and models to simulate the impact of large construction works, traffic flows and service provisions; and games developers looking to build entirely new worlds from scratch. All these applications use technology to make new mappings of the world, but how do they shape the way *we* see the world?

Architects and property developers have been especially keen to use geo-visualizations, which include maps, models and computer-generated imagery, to show off their designs and bid for new contracts. In some countries it has become a mandatory part of the tendering process, alongside architectural drawings and costings, for large civic projects. This is because highly accurate digital images of how a design will look *in situ* are very effective at selling an idea, and not just a building, to clients.[8] These imaginary mappings, which represent a world yet to emerge, are found on construction hoardings, project websites and polypropylene boards at public consultations. They depict perfectly formed, blemish-free atmospheric futures of worlds that are somehow connected to and yet clearly detached from our own.[9] Buildings are glossed, infrastructure is well maintained, foliage is trimmed and transport appears to run smoothly, while idealized occupants (if they are even included) are plucked from databases and placed strategically into the digital landscape as what the artist James Bridle has described as 'render ghosts'.[10]

Geo-visualizations are not strictly maps in the conventional sense of a two-dimensional representation. However, they do have map-like qualities, including the representation of scale, topography and human geography. In versions of the technology used alongside the marketing material, these links are even more apparent. Building Information Modelling (BIM) is a standardized three-dimensional model and data-set initiative designed to simplify how geographical data about a site is stored and shared among contractors involved in the planning, construction and maintenance of a building over its life cycle. It works by sharing a single geographically precise model with all the contractors involved, including architects, engineers, builders and utilities providers. It is said that BIM will transform the industry by streamlining the antiquated system whereby each stakeholder in the process must do their own survey, design, data collecting and ground truthing before they can begin work on site.

In the highly competitive practice of bidding for urban planning contracts, architects can spend great sums on modelling software and design expertise in order to develop the best possible imagery.

It is estimated that by 2028 the market for three-dimensional visualizations will be worth almost $7.5 billion. NEOM, an urban-planning consortium led by the crown prince and prime minister of Saudi Arabia, Mohammed bin Salman, has taken this practice to new extremes with its aim to build an entirely new city from scratch

top: Digital rendering of an imagined science park, aerial view, 2023.
bottom: Digital rendering of an imagined cityscape, pedestrian view, 2023.

in the Saudi Arabian desert. Launched in 2017, The Line is NEOM's as yet unbuilt vision for the future of urban living based on the idea that a condensed linear design can help to reduce carbon footprints while creating the conditions for better social relations. It aims to be just 200 metres (656 ft) wide and 170 kilometres (106 mi.) long, to accommodate 9 million people in a space measuring 34 square kilometres (just over 13 sq. mi.), to have no roads and to be run entirely on sustainable energy.

These lofty goals are illustrated in the geo-visualizations that make up much of the marketing material. In one promotional video we follow the dream-like experience of a potential resident who is transported from their drab urban existence to a vibrant, multi-sensory future in the desert. The viewer is led on a transformative journey to believe that another version of urban life is possible, one characterized by new forms of social and human–nature relations, and encapsulated in the tagline 'New Wonders for the World'. It is pure marketing, bordering on the preposterous, and few would bet on this exact vision becoming a reality by the stated completion date of 2030, but it does show how effective geo-visualizations can be in selling an idea, which is often what matters when bidding for investment and attracting future residents.

More like a computer game or science-fiction movie made with CGI, NEOM's marketing follows a trend of combining maps, models, plans, illustrations and computer-assisted design in order to sell ideas about urban living that reimagine the city from scratch. The Line certainly echoes the fictional futurism inspired by such popular games as *SimCity* and *Townscaper*, in which gamers can play God and map out their urban vision from the comfort of their homes, but there is also a real history that dates back centuries: of urban planners building models, maps and sketches to inspire, instruct and compete for attention.[11]

Today, from the hallways and exhibition spaces of decadent urban-design expos around the world to the walls of local councils' planning meetings and the screens of design studios, ideas about urban living are being mapped out in high-definition geo-visualizations. And in the future, we can expect more technology like this to bend our conception of what a map is. The idea that a map must be a two-dimensional image from above has already been superseded as we have become used to three-dimensional projections, heads-up displays, toggled overlays and real-time updates. As the technology advances further, we can expect our ideas of the map to blur even more, perhaps to the point where conventional definitions of the map no longer make sense.

*

At the same time that self-driving cars and geo-visualization technology are mapping out new worlds and ways of seeing, machine learning is changing what it means for us to read the maps we already have. For navigational purposes, machine-readable maps have been used for some time to identify changes on roads, streets and paths, and to calculate more efficient routes, whether by car, bike or public transport, or on foot. The up-to-date maps and directions that we have become accustomed to every time we load a route on our phones rely on this type of technology, which is designed to recognize changes in the world for us. The likes of Google Maps and its competitors collect such large amounts of geographical data each day – ranging from mapped data to location databases and photographic images – that it has become necessary for machines to wade through it and identify the changes within it, if it is to be of any use.[12] Once a machine has been trained to identify what to look for – common building types, road surfaces, layouts, street furniture, traffic lights, signage and more – map-reading and routing at scale and in quick succession become remarkably more efficient and cost-effective than if they were to be done by a trained human analyst.

Across sectors ranging from cultural heritage and conservation to insurance, risk analysis and even agriculture, similar machines are being trained to read maps, extract and classify the information and objects found within them, and link this with other maps and databases to understand how people and places change over time. The Machines Reading Maps project, a collaboration by the University of Southern California Digital Library, University of Minnesota Computer Science and Engineering Department and Alan Turing Institute in London, has done just that, by creating a tool to search for text in historical maps, just as you would search for text in a digital document. Since August 2023, users have been invited to read maps in a fundamentally new way, by searching across maps from different periods and type, and with other databases, all on one web-page. This shifts the painstaking task of searching for details on maps into something that can be done in seconds, as machine-reading technology is put to work identifying and extracting text from digitalized maps. For instance, users can now search for all 'public houses' or 'police stations' that appear in the database of maps, made up from Ordnance Survey collections from the British Library and the National Library of Scotland, and maps from all over the world held by the David Rumsey Map Collection.

These so-called deep mapping techniques will open the door to new ways of interrogating maps and create new pathways between maps. Historians and heritage professionals could benefit from the technology because they will be able to study how the cartographic record of places has changed over time, which in turn could help them uncover new archaeological sites, update historical perspectives and gain further insight into how people lived their lives. Consider the city of London, which has been mapped for centuries. With this technology, users could examine the changes in the city by carrying out searches for text in some of the earliest known maps, from the Agas map of 1561 to the Great Fire of London map from 1667, and right the way through to twentieth-century maps, including the famous A–Z street map first made in 1936, and the bomb-damage maps produced between 1940 and 1945 to show the damage caused by the Blitz.

The management of contemporary urban growth has already benefited from this technology. In Zambia, the local government of the capital city, Lusaka, has worked with the Ordnance Survey in the United Kingdom to produce a map of the city's growing footprint by deploying a machine-learning tool trained to detect buildings and informal settlements from aerial, drone and satellite images. The resulting base map covers 420 square kilometres (162 sq. mi.) and has become a valuable planning tool to help the government understand where urban sprawl is happening and how best to manage it. An up-to-date map of the growing city can be used to plot public-transport routes, locate social services and plan where to build infrastructure projects. The base map is also a proof of concept that shows how cities can be automatically mapped from afar, for relatively little money and at speed. We can expect similar technology to be replicated in under-mapped cities across the continent in the years to come.[13] It will be used by not only city planners, but commercial ventures as they try to map new markets.

In Uganda, where a crisis is unfolding as the country grapples with the task of accommodating 1.5 million refugees who have settled there after fleeing violence and food insecurity mostly from South Sudan and the Democratic Republic of Congo, this same technology is being adopted by humanitarian organizations.[14] Microsoft's AI for Humanity initiative is working with the Humanitarian OpenStreetMap Team (HOT), Bing Maps and the Netherlands Red Cross to develop map-reading technology that automatically identifies informal migrant settlements from aerial images, to speed up the process of making up-to-date maps available to humanitarian aid operations. For workers on the ground, maps that show where migrant populations are settling are an especially important way to view the scale of the problem and decide where to direct their attention. And although human volunteers are integral to the process of ensuring the maps meet quality-control standards, machine-learning technology does offer a clear advantage in speeding up the process, which matters if lives and livelihoods are on the line.

What is interesting about this particular case is the context in which the technology has been deployed. HOT is a community built from hands-on contributions from thousands of volunteers, often working together in a social environment to update the map. Automating the processes involved might disrupt the social nature of the project, and as such HOT has been careful to survey the community about where and when to use the technology. It found that contributors saw clear benefits to automation speeding up the mapping process, but also that the community was not always willing to trust machine learning to produce maps of the highest quality, which is where they saw value in their own expertise. This is not usually seen in other uses of the technology, where fully automated mapmaking is being done with little regard for the human mapmakers involved.

Programming machines to read maps can lead to breakthroughs in how we see the world. Using this technology to map and track the impact of social and environmental change at scale does provide a clear overview of what is happening, which is useful for science, industry, policymaking and humanitarian efforts. But this is a powerful technology that, if left unchecked, can lead to significant problems. There is evidence to suggest that misuse, abuse and unintentional consequences of relying entirely on machines to make sense of the world for us are common. As the HOT case shows, we should take a cautious approach to mapping at scale and at speed. It is one thing to produce maps in this way to prove it can be done, and another to consider the wider implications of doing so. What does it mean to produce a map automatically from afar? To trust a machine to lay out the lives of people without ever coming into contact with them, or without them even knowing? Such questions have long been debated in critiques of mapmaking, but they come to the fore in ever more apparent ways when we give machines free rein to make decisions over what aerial images show, decisions that could directly affect the lives of people on the ground.[15]

*

Outside specific industries and academic disciplines, most people have a limited understanding of how self-driving cars make decisions about what to do next, or how machines perform the task of identifying features on a map. All we are left with is the map that emerges from this process, and any explanation the team behind it can give. Much of the time this does not matter – we do not need to know how a system works in order to benefit from it – but what mapping technology does is add further layers of abstraction between us and the world, meaning that our interpretation of it is shaped further by views from above and by technology that is difficult to understand.[16] What we see helps us to conceptualize the world and the scale of its problems, but there is still some distance between what these maps show and how we understand lived experiences on the ground.

So, what can be done? We could look to alternative mapping technology that foregrounds human experience, to complement advances in machine learning. One area that will surely receive interest in the coming years is machine-readable maps that help to identify the impact of climate change on humans and the planet.[17] In this context, maps and machine-learning algorithms that are more likely to be designed by computer scientists than classically trained cartographers offer a way to locate, track, monitor and manage environmental phenomena automatically, at a global scale. This makes them useful for studying the effects of climate change across space and time, and influential in steering policy towards mitigation and sustainability targets. The underlying principles of such mapping technology are similar to those we have discussed; they use computers trained on aerial imagery and remote-sensing data collected by drones, planes and satellites to classify objects on the ground, record changes over time and predict what might happen next. Early iterations of this technology have already been used to identify and predict threats of deforestation, as well as to monitor glacial melting and locate the areas most susceptible to drought, rising sea level and flooding.

In this context, the maps produced will provide scientific insights into the scale of the problem, but, as recent history shows, climate change and its impact are not scientific problems to be solved by the neat offering of location data plotted on aerial-view maps. So-called 'wicked problems' are the main impact: problems where social, cultural, political, economic and environmental elements intersect. These are problems that transcend gridlines and escape cartographic convention, requiring us to think again about how we approach them with maps.

Story maps are an alternative that we already have.[18] These are maps that link places with personal narratives and collective histories of the people who live and have lived there. They add a qualitative dimension to maps that brings them to life and helps to illustrate the complexity of interconnected problems. This is not something most maps do today, even though using maps to tell and represent stories is a practice that dates back centuries. Nevertheless, with the advent of digital mapping technology, easy-to-use web-based platforms and a growing interest in spatial storytelling, there has been a revival in putting stories on to maps and maps into stories. This has been done to show a human side to global problems, such as climate change, but also as a narrative device for storytelling.[19]

The crowdsourced Corona Diaries project made during the COVID-19 pandemic is one example of how wicked problems can be mapped with a different, more human-centred approach. This story-mapping project gives participants the opportunity to locate and share their experiences of lockdowns around the world by uploading audio recordings that are pinned to the map. A personal connection is created as these stories – told in the voices of the contributors – are shared. The testimonies highlight the sheer range of experiences people had, despite the commonality of being locked down in their homes. There are tales of peaceful solitude and the renewed appreciation of local spaces, mixed with those of tense family relations, social claustrophobia and growing anxiety about humans' impact on the world.

Corona Diaries offers a view on to how people live through wicked problems, and in this sense it does much more than a standard cartographic map could to highlight the messiness of issues that are all too often packaged in neat ways. But story maps are by no means perfect. The geographers Sébastien Caquard and Stefanie Dimitrovas suggest that the complexity of narrative will always be difficult, if not impossible, to represent on maps when they say that 'maps and stories simply do not have the same geography.'[20] Stories are tellings of lived experiences that transcend fixed spatial and temporal boundaries, they continue: 'Transforming stories into maps is particularly challenging due to the tension between the blurry, personal, and emotional dimensions of stories and the characteristics of fixity, hierarchy, and quantification inherent in conventional cartographic representations.'[21]

This makes stories a real challenge for maps, where people and places are fixed in space by design. Narratives aside, web-based versions also offer only those with the necessary technical skill the opportunity to contribute, and they would require significant start-up and running costs should you want to create your own from scratch. However, they do point to a mapping future that embraces complexity in ways that many of us can understand. This is already doing more than comparable maps of wicked problems, which continue to leave the human stories off the map and treat these difficulties as purely scientific issues to be solved.

Story maps put human voices on the map, but we could go further still and reconsider whose spatial knowledge we centre in our fight against globally shared wicked problems. Much like the pandemic, climate change is interconnected locally and globally, yet we tend to neglect the local and so-called non-scientific in favour of a globalized and 'scientific' view, favouring objective facts over lived experience, as if a focus on the science at scale is the only way to get to grips with the problem. One way we might challenge or complement this view is by listening to people who have a deep knowledge of places and how their ecology has been affected. This requires that a different form of knowledge-making be recognized as legitimate. It could also mean resisting the temptation of such abstractive mapping technology, and taking seriously other mappings that challenge cartographic convention.

Listening to life on the Earth and hearing what it needs to survive is not always something we can do in the abstract. Perhaps it never is. No matter how sophisticated our cartographic technology, it always takes us some way from the lived experiences of being on the Earth: touching it, reading it, smelling it, seeing it, sensing it.[22] We cannot live on the cartographic map, after all. But there are other mappings of the world that can live within us – living maps, perhaps – spatial knowledge that we can use to make sense of the world's problems and how we might address them.

Indigenous communities around the globe have, for a long time, been the custodians of living maps that emphasize lived and multi-sensory forms of relational engagement with the Earth. These are maps that live in the mind more than they do on paper or screen, mappings that understand that what happens in one place has an impact on what happens in another. This is spatial knowledge that sees little use in placing abstract borders over ecological processes that do not abide by them (even though this remains a standard practice of land management). Indigenous knowledge is usually sacred and not for sharing widely. For very good reason, too: its past sharing with colonizer settlers has led to all manner of cultural and ecological destruction. However, a common principle of these living maps that we can learn from is to think of maps through a framework of lived experience and ecological relations, rather than through a framework of abstraction. This offers a valuable lesson for our mapping futures.

There is already traction here, as many Indigenous scientists and activists make calls to think about mapping the Earth with Indigenous knowledge frameworks in mind. Hindou Oumarou Ibrahim, an award-winning activist, is a notable advocate of this approach. Working with the Association for Indigenous Women and Peoples of Chad, and Mbororo elders from around Lake Chad, Ibrahim produces participatory maps that help to mitigate the impact of climate change on nomadic farming communities in the region. Combing satellite imagery with the elders' knowledge of land and sky, these maps are used to show geographical features, such as the location of fresh water that can be used for livestock and crop irrigation, land corridors most suited to pastoral activity and what happens to land and water supplies in the areas most susceptible to adverse weather. These maps are made for pragmatic reasons, as a response to the lack of meteorological forecasting provided to these communities, and for negotiating purposes, so that farmers can agree on local access rights to the land. But equally they demonstrate the value of using an Indigenous knowledge framework over a purely Western-scientific one, to shape the process of making climate-mitigation maps.

It remains to be seen whether Indigenous knowledge will be taken seriously at scale in mapping projects that aim to fight climate change. A great deal of lip service is still being paid to Indigenous knowledge, and there is much evidence to suggest that the value of this knowledge is not being repaid with acts that benefit these communities. But over the past two decades there has been a notable positive shift, owing to the work of Indigenous people like Ibrahim, who make huge efforts to circulate these ideas in society.

The idea of thinking with Indigenous knowledge frameworks in mind does not mean all people must return to the Earth as farmers and somehow reverse developments in society – a common criticism of calls to think again about the current state of play – nor do they mean to suggest that we all appropriate Indigenous culture as our own. As discussed elsewhere in this book, Indigenous knowledge is often not something that can be simply shared; there are protocols to follow, often designed specifically so that this knowledge is not misused. Rather, this is a call to think again about our relationship with our life-support system – the Earth – and to recognize that there are alternative mapping futures that we could learn from, if we took the time to listen and respect them.

We started this chapter with a look at how digital technology is mapping the world in ever more precise detail, with the promise of giving us and machines an ever greater understanding of the world. This may well be true, but I am reminded of the idea that the closer you look, the more distant you become. It may well be that despite all the benefits of the cartographic technology in use today, we are losing touch with the Earth, and with other ways of mapping it that may be more useful in helping us understand it. At the very least, our largely scientific mapping of the world would benefit from acknowledging and respecting the value of living maps. Reflecting on the counter-maps made by Jim Enote (see Chapter Two), the writer Chelsea Steinauer-Scudder puts it this way:

> The Zuni maps remind us that modern, conventional maps convey only one very particular way of being in place, one which often, counterintuitively, leaves us disoriented and disconnected. Conventional maps do not tell us what it means to be somewhere – the details of the landscapes we live in, the sounds of the trees and the birds, the long histories of the arroyos and the mountains, the names of the people who built our homes.[23]

Where does this leave us when thinking of our mapping futures? From the stories told in this book we can begin to understand that mapping is not on a linear path of progress or technological development. Rather, we have seen that it is always a social, cultural, political and economic process that acts according to its own divergent rules, which do not always have a clear logic.

Maps are in a constant state of iteration and are *said* to become more advanced, detailed and objective as each year passes. Throughout history there have been periods of major progress, each claiming to offer a revolutionary leap in how humans see the world from afar. This includes the establishment of cartographic standards in the eighteenth century, the development of printing technology, aerial photography and satellite imagery in the twentieth century, the emergence of digital Geographic Information Systems (GIS) in the 1960s, the opening of GPS to the public in 2000, and the rise of online mapping platforms, massive location-based data sets, and maps on mobile phones. Most recently these developments can be marked by the deployment of sensor technology and machine-learning algorithms that read and make maps for us, and the burgeoning use of geo-visualization technology that puts maps alongside three-dimensional models and other computer-generated imagery to reshape our understanding of what maps are for.

Bound up in this progress has been a host of new terminology. 'Maps', 'cartography' and even 'GIS' are no longer enough to describe and define the range of ways that geographic information is visualized and presented in society today. Now we have 'geo-visualizations', 'geo-spatial technology', 'locative media', 'spatial media', 'neogeography' and 'machine vision' to contend with.[24] Each has a specific meaning and is tied to particular fields and uses. They include elements of what we might see as conventional mapping, but each term extends those conventions with technology, and stretches the way the world's geography can be represented and understood. These terms are largely used by academia and industry, and are unlikely to replace 'maps' in the general lexicon anytime soon. But they do point to the diversity of ways that mapping has been understood in recent decades, and perhaps towards a future in which our definitions of the map no longer make much sense.

Making and using maps is also a conceptual and philosophical practice that evolves alongside mapping technology, and sometimes without any regard for it. The interest in 'sensory maps' is a case in point. Smell maps and sound maps are growing in popularity as a means of representing the spatiality of sensory experience, as well as an invitation for us to think beyond the idea that maps are devices for showing only visible phenomena. For example, Kate McLean's smell maps of cities around the world are illustrative of how experiences of urban places are shaped by the temporality of the smells encountered. These maps, which blend the nuances of olfactory experience with static and sometimes dynamic representations of space, point to future directions in mapmaking that are seemingly far removed from those seeking to develop technology that maps the world for automated decision-making. Sound maps, too, like those supported by the CGeomap platform, are becoming a very popular means to rethink what advances in mapping technology could and should be used for. David Merleau is one artist on the platform experimenting with how to map sound. White Bear Trail is his location-based sound map that guides the listener along a forest trail outside Temagami in Ontario, Canada. As the listener moves through space, whether that be the digitally mediated space of the screen or the physical space of the forest, their journey is interspersed with sounds and stories from the forest, designed to augment a cartographic practice with a sensory experience.

These divergences in mapping futures can be seen most clearly in the humanities and the arts, as well as within the Indigenous knowledge frameworks discussed earlier. Often, they come from people who do not have a background in making maps. They point to a different future for maps: one that grapples with the technology *and* the concept of mapping.

For as long as we find value in using maps to represent the world, we can expect this rhizomatic trajectory of development to continue, with each new advance in technology and conceptual framework claiming to surpass the last. There will always be new ways of making maps, and new ways of using maps. Regardless of the technology, we can be sure that any new maps that go out into the world will become entangled very quickly in all kinds of social life. We see this at the current juncture, when sophisticated machine-mapping technology for self-driving cars is confronted by the reality of complex driving scenarios, legal systems and popular car culture. And we see this in the deployment of geo-visualization technology in urban design, which produces dazzling images of the future but cannot be separated from the reality of planning laws, construction constraints, community pushback and high financial cost. This follows those who developed early GPS mapping technology for the U.S. military in the 1970s, who could not have foreseen how this would come to shape social life, and those who developed the tools for the mass production of maps in the early twentieth century, who could not have predicted the many things that people have done with paper maps ever since.

Maps are born into and become part of the world(s) they aim to represent, in turn becoming part of the lived experiences of those people who encounter them. As we have seen, maps are produced in social contexts rather than from neutral spaces. These bonds only become tighter as we adopt them within our ways of life. All we can do as people interested in maps is pick away at and untangle a few of these threads to understand what is really going on. Understanding the context in which we use maps still matters if we truly want to judge the impact they have on society, regardless of the technology.

Despite the ways technology will shape the production, use and circulation of maps in the years to come, our future with maps is likely to remain a messy business, where users will continue to develop their own mapping practices based on what suits them in any given moment. In this sense, I do not see an end to users combining paper maps, printouts, smartphones and satnavs for navigation. Nor do I envision a future where we are all using maps in the ways that they were designed for, or a time when the definition, meaning and value of maps suddenly become stable. I only see more maps being added to this mix, and more mapping practices developing as a result. This is because we shape maps at the same time as they shape us. And, as we have seen, class, culture, politics, economics, education, gender, race, environment and technology all have the power to shape how and why we use maps in a particular context.

Rather than squeezing the map or user into a neat, predefined box, we could take more seriously the diversity of ways that maps enter and shape people's lives and livelihoods. This would tell us not just about maps, but also about how people view, understand and act in the world with maps. This is not just an academic exercise, either. Recognizing these differences is important for mapmaking, because it will drive us towards producing maps that better represent the diversity of world views, and away from maps that claim to present a universal truth. We have seen this diversity being reflected in counter-mapping projects and through the work of amateur mappers made possible by the Internet and digital mapping technology, but those who dominate the profession of cartography have not yet taken up the challenge to any great extent, largely because it would disrupt established design principles, damage business models and call into question the perception of objectivity that has been so useful for those with the power to wield it.

Counter to the claims that some maps have changed the world, I argue that *all* maps have changed the world and have the capacity to continue to shape it no matter how new, sophisticated, old or outdated they might be. The question is: whose world have they changed, and how are these changes noticed, written about and turned into narratives that are circulated in society? There are millions of maps in this world, and very little is known about how they shape the people who live in it.

To make these arguments, I have focused on how maps are used for navigation and movement tracking, and how they are used to represent and shape politics, culture and the economy at different scales. In each chapter I have looked into different situations, exploring various mapmakers and users to highlight how the context of a map's production, use and circulation matters. In this chapter, I have added a final dimension to my argument by focusing on where we go from here, at a time when the digitalization of mapping practices continues at pace and when the scientific standardization of cartographic maps is ushering out other ways of mapping the world and humans' relationship to it. Dominant knowledge of mapping presents is used to frame the discourse on mapping futures, but there are alternative mappings that could offer different ways to approach even our most pressing concerns. How do we incorporate such marginalized, but no less expert, knowledge into planning for the future? What kind of future with maps do we have if we prioritize certain mapping knowledge over others?

Epilogue

Maps are geographical representations with deep social, cultural, political and economic significance. They will always be objects of geographical knowledge, but they do much more to shape our daily lives than we might first expect of a representation of the Earth. Professions that produce maps of the highest quality, and fields of research that study maps in minute detail, are well established. This work can be seen in the maps we use every day to get around, and in those used by experts and specialized industries, and how they are depicted on the pages and screens of popular culture. In stark contrast, there is surprisingly little interest in how or why people use maps, and how a map's use is shaped by the context of that use. This book is my attempt to push the conversation about maps in that direction, to highlight maps of many kinds and the diverse situations in which they are used, with the aim of developing a greater recognition of how maps shape the world.

To conclude, I offer a set of provocative questions to ask yourself next time you pick up a map. These are neither exhaustive nor necessarily useful for everyone, but they are designed to engage you with the arguments presented in this book. They are the questions that I ask of any map, be it digital, paper, rock carving or folk story, and questions to be asked of any style of mapping, be that maps produced using the latest cartographic techniques or those loosely sketched with a pencil or pen. They are questions that you can ask others, too, on first using a map, after repeated uses, or as that map collects real or metaphorical dust. I hope they will be as useful for you as they have been for me in thinking about the work maps do in the world.

What is this map designed to be used for, and why did I choose to use it?

How does this map feel to touch and use?

Does it stir any memories or emotions, or none at all?

Whose culture and politics are represented on this map?

Who might benefit from what is on the map, and who might lose out?

Do I value this map? Would someone else value it in the same way?

Do I know why or how this map was made?

What will happen to this map once I am finished with it?

REFERENCES

Introduction

1 Matthew Edney, *Cartography: The Ideal and Its History* (Chicago, IL, 2019).
2 See ibid. and the History of Cartography Project, University of Chicago; Lucia Nuti, 'The Perspective Plan in the Sixteenth Century: The Invention of a Representational Language', *Art Bulletin*, LXXVI (1994), pp. 105–28.
3 For an introduction to the history and etymology of 'cartography', see Peter van der Krogt, 'The Origin of the Word "Cartography"', *e-Perimetron*, X/3 (2015), pp. 124–42.
4 Edney, *Cartography*.
5 Emanuela Casti, 'Bedolina: Map or Tridimensional Model?', *Cartographica*, LIII/1 (2018), p. 26.
6 Catherine Delano-Smith, 'Cartography in the Prehistoric Period in the Old World: Europe, the Middle East, and North Africa', in *The History of Cartography Project*, vol. I, ed. J. B. Harley and David Woodward (Chicago, IL, 1987), pp. 54–102.
7 Ibid., p. 80.
8 I have decided to talk about the spatial knowledge and mapping practices of this group and the following Pacific Islanders in relatively general, rather than specific, terms, using research produced by Indigenous scholars and allies. This means not reproducing names, stories or specific details about the land relations that are a key part of these cultures. I have no right to write in specific terms about the knowledge and practices of individual Indigenous communities without seeking permission from them. This is not my knowledge to share. Nevertheless, the authors whom I cite do have the right to discuss these details, and it is their work that should be read for a deeper understanding. I am aware that this is at odds with how knowledge is shared in non-Indigenous publishing contexts.

Much of my discussion here is inspired by the Gay'wu Group of Women, who write specifically about the mapping practices of Indigenous people living in Yolŋu Country, in northeastern Arnhem Land, Australia, but there are many other communities that have their own mapping practices. Please do not take what is written here as a definitive guide to how songlines work across the Indigenous peoples of Australia. See Gay'wu Group of Women, *Songspirals: Sharing Women's Wisdom of Country through Songlines* (Sydney, 2020).

9 There is no one way in which Pacific Islanders use these environmental cues for navigation. The area is vast and has its own geography of navigational knowledge practices.

10 As Fetaui Iosefo, Anne Harris and Stacy Holman Jones have argued in *Wayfinding and Critical Autoethnography* (Abingdon, 2021), the academic translation of 'navigation' or 'wayfaring' to 'wayfinding' has a tendency to decontextualize and disrespect these Indigenous practices, ultimately leading to a misunderstanding of what they are actually about.

11 Ibid., p. 21.

12 A point of tension in writing this book has been how to balance my analysis of maps and mapping practices by looking around the world with respecting the fact that this knowledge is not always mine to share. I could have, and perhaps should have, chosen to exclude this section, but doing so didn't seem right, since it might have further excluded and othered this knowledge from the discussion about maps, which remains stubbornly focused on Western topographic maps and mapping practices. I hope what I've written here both introduces a wider audience to different forms of maps and mapping practices and, more importantly, encourages them to engage with Indigenous experts on the topic, who know far more than I do. I am still learning about how to write about and navigate this tension in my work on maps, and what I have written will almost certainly not satisfy everyone.

13 Akiemi Glenn, 'Wayfinding in Pacific Linguascapes: Negotiating Tokelau Linguistic Identities in Hawai'i', PhD thesis, University of Hawai'i, 2012; Iosefo et al., *Wayfinding*; Klara Kelley and Harris Francis, 'Traditional Navajo Maps and Wayfinding', *American Indian Culture and Research Journal*, xxix/2 (2005), pp. 85–111; Pamela Colorado, 'Wayfinding and the New Sun: Indigenous Science in the Modern World', *Noetic Sciences Review* (Summer 1992), pp. 19–23.

14 'Navigational technology' has commonly become associated with modern, digital and Western technology, but we should also consider long-standing, non-digital and non-Western technology, such as the stick-and-shell charts used by Polynesian wayfinders, as modern

technology. Not to do so perpetuates the myth that Indigenous navigational practices are somehow stuck in the past, which they are not.

15 This tweet has since been deleted, owing to Twitter blocking Trump's account in January 2021 (it was reinstated in November 2022), but it is widely available online.

16 See www.elephantandcastle.org.uk/elephant-and-castle-regeneration-map, accessed 1 February 2023.

17 Campaigns to highlight and fight the displacement of Latin-American and other communities have developed in response to the gentrification of the area. See, for example, the Up the Elephant campaign.

18 Maps rarely exist in isolation. Often they are placed alongside other images, text and data. Reading the context of maps requires that attention be paid to them in relation to the material alongside which they sit, and the situations in which they are used. There is a strand of cartographic theory that uses what is known as 'actor network theory' to make sense of these relations. See Bruno Latour, *Science in Action* (Cambridge, MA, 1987); Mike Duggan, 'Cruising Landscape-Objects: Inland Waterway Guidebooks and Wayfinding *with* Them', *Cultural Geographies*, XXIX/2 (2021), pp. 167–83; Veronica della Dora, 'Performative Atlases: Memory, Materiality, and (Co-)Authorship', *Cartographica*, XLIV/4 (2009), pp. 240–55.

19 The aestheticization of gentrification and displacement is a long-standing practice. See, for example, Anastasia Baginski and Chris Malcolm, 'Gentrification and the Aesthetics of Displacement', FIELD, 14 (Autumn 2019); Christoph Lindner and Gerard F. Sandoval, *Aesthetics of Gentrification: Seductive Spaces and Exclusive Communities in the Neoliberal City* (Amsterdam, 2021).

20 Shelley Rice, *Parisian Views* (Boston, MA, 1999).

21 Mike Duggan, 'Mapping Interfaces: An Ethnography of Everyday Digital Mapping Practices', PhD thesis, Royal Holloway, University of London, 2017.

22 Martin Dodge, Rob Kitchin and Chris Perkins, ed., *Rethinking Maps: New Frontiers in Cartographic Theory* (Abingdon, 2009), p. 5.

23 Greg Milner, *Pinpoint: How GPS Is Changing Technology, Culture, and Our Minds* (London, 2017); Matthew W. Wilson, *New Lines: Critical GIS and the Trouble of the Map* (Minneapolis, MN, 2017); Rob Kitchin, Tracey P. Lauriault and Matthew W. Wilson, ed., *Understanding Spatial Media* (London, 2017).

24 Duggan, 'Mapping Interfaces'.

25 See, for example, Clancy Wilmott, *Mobile Mapping: Space, Cartography and the Digital* (Amsterdam, 2020).

26 Ibid.; Duggan, 'Mapping Interfaces'.

27 There is a growing body of research in the neurosciences and geography that is studying the impact of GPS on people's ability to navigate. For example, Janet Speake and Stephen Axon, 'I Never Use "Maps" Anymore: Engaging with Sat Nav Technologies and the Implications for Cartographic Literacy and Spatial Awareness', *Cartographic Journal*, XLIX/2 (2012), pp. 326–36; Louisa Dahmani and Véronique D. Bohbot, 'Habitual Use of GPS Negatively Impacts Spatial Memory during Self-Guided Navigation', *Scientific Reports*, X/1 (2020), article no. 6310; Pablo Abend and Francis Harvey, 'Maps as Geomedial Action Spaces: Considering the Shift from Logocentric to Egocentric Engagements', *GeoJournal*, LXXXII/1 (2017), pp. 171–83. There is also a strand of journalism that has become interested in the topic. See, for example, David Kushner, 'Is Your GPS Scrambling Your Brain?', *Outside*, www.outsideonline.com, 15 November 2016. This is interesting work, but it often relies on the assumption that GPS has made us somehow unaware of our surroundings. Clearly this is untrue, or people would be bumping into things all the time and no one would ever know where they were. I am willing to bet that you knew something of the places you passed through the last time you used your satnav. Running parallel to this are two more bodies of scholarly work. The first demonstrates how the impact of GPS on the brain should not be divorced from the social act of navigation, in which who we are with and what we are doing often shapes our navigational abilities, too. See, for example, Eric Laurier and Barry Brown, 'The Normal, Natural Troubles of Driving with GPS', *CHI'12 Proceedings: Proceedings of the SIGCHI Conference on Human Factors in Computing Systems* (Austin, TX, 2012); Mike Duggan, 'Navigational Mapping Practices: Context, Politics, Data', *Westminster Papers in Communication*, XIII/2 (2018), pp. 31–45. The second asks us to consider how the navigation of space is, in the digital age, always already embedded with digital ways of seeing and socializing. See, for example, Alex Gekker and Sam Hind, 'Outsmarting Traffic, Together: Driving as Social Navigation', in *Playful Mapping in the Digital Age*, ed. Playful Mapping Collective (Amsterdam, 2016), pp. 78–93; and Sam Hind, 'Digital Navigation and the Driving-Machine: Supervision, Calculation, Optimization, and Recognition', *Mobilities*, XIV/4 (2019), pp. 401–17.

28 See the work of Agnieszka Leszczynski, who ties a political discussion about maps and spatial data into that of platform urbanism, the notion that platforms are increasingly shaping and dominating urban

experiences and practices. See 'Platform Affects of Geolocation',
Geoforum, cvii (2019), pp. 207–15, and 'On the Neo in Neogeography',
Annals of the Association of American Geographers, civ/1 (2014),
pp. 60–79.

29 See also Mike Duggan, 'The Everyday Reality of a Digitalizing
World', in *Geographies of Digital Culture*, ed. Karsten Gäbler and Tilo
Felgenhauer (London, 2017), pp. 71–83.

30 Duggan, 'Mapping Interfaces'.

31 There exists an industry around this idea since it first came about
in Brian Harley's now infamous paper on 'deconstructing' the map
in 1989. See J. B. Harley, 'Deconstructing the Map', *Cartographica*,
xxvi/2 (1989), pp. 1–20. These ideas have been taken up in the field of
critical and historical cartography as much as in the pages of popular
non-fiction books, television documentaries and well-publicized
exhibitions about maps.

32 Maps as metaphors are something different. They may be used to
represent spatial elements, but most are used to conceptualize an
overview of related elements, that is, to *map out* a field of research.

33 Denis Cosgrove, ed., *Mappings* (London, 1999), p. 2 (emphasis original).

34 See, for example, Jeremy Crampton, *The Political Mapping of Cyberspace*
(Edinburgh, 2003); Jeremy Crampton, 'Cartography: Performative,
Participatory, Political', *Progress in Human Geography*, xxxiii/6
(2009), pp. 840–48; Dodge et al., *Rethinking Maps*; Wilmott, *Mobile
Mapping*; Tania Rossetto, *Object-Oriented Cartography: Maps as
Things* (Abingdon, 2019); Les Roberts, *Mapping Cultures: Place,
Practice, Performance* (London, 2012); John Pickles, *A History of Spaces:
Cartographic Reason, Mapping and the Geo-Coded World* (Abingdon,
2004); James Corner, 'The Agency of Mapping: Speculation, Critique
and Invention', in *Mappings*, ed. Cosgrove, pp. 89–101.

35 As Denis Wood has suggested, maps have an ability to convince us
of their simplicity (and neutrality) because, more often than not, they
work for us. In other words, we don't often see beyond their use as
a functional device. See Denis Wood, *Rethinking the Power of Maps*
(New York, 2010).

36 For a detailed analysis of the power of maps as things in and of
themselves, see Rossetto, *Object-Oriented Cartography*.

1 Navigation Beyond the Map

1 See J. O'Keefe and J. Dostrovsky, 'The Hippocampus as a Spatial Map:
Preliminary Evidence from Unit Activity in the Freely-Moving Rat',
Brain Research, xxxiv/1 (1971), pp. 171–5; Russell A. Epstein et al.,

'The Cognitive Map in Humans: Spatial Navigation and Beyond',
Nature Neuroscience, xx (2017), pp. 1504–13; M. R. O'Connor,
*Wayfinding: The Science and Mystery of How Humans Navigate
the World* (New York, 2018).

2 See Eleanor A. Maguire, Katherine Woollett and Hugo J. Spiers,
'London Taxi Drivers and Bus Drivers: A Structural MRI and
Neuropsychological Analysis', *Hippocampus*, xvi/12 (2006),
pp. 1091–101; Katherine Woollett and Eleanor A. Maguire,
'Acquiring "The Knowledge" of London's Layout Drives
Structural Brain Changes', *Current Biology*, xxi (2017), pp. 2109–14.

3 For an engaging introduction to the Knowledge, see Robert Lordan,
The Knowledge: Train Your Brain Like a London Cabbie (London, 2018),
but for detailed information on what the Knowledge entails, see the
so-called Blue Books given to Knowledge students as they begin their
learning. The official name of said Blue Books is the *Guide to Learning
the Knowledge of London*. Published by Transport for London, they are
updated every year.

4 Nevertheless, there are examples of when efficient orienteering with a
map has become a way of life, such as within geocaching communities,
Scout clubs around the world, the military and various obscure
orienteering events that attract large numbers of dedicated followers
(for example, the annual O-Ringen event in Sweden).

5 There is a long history of navigational devices being used in cars.
See, for example, Tristen Thielmann, 'Linked Photography:
A Praxeological Analysis of Augmented Reality Navigation in the
Early Twentieth Century', in *Digital Photography and Everyday Life:
Empirical Studies on Material Visual Practices*, ed. Edgar Gómez
Cruz and Asko Lehmuskallio (London, 2016), pp. 160–85; James
R. Akerman, *Cartographies of Travel and Navigation* (Chicago, IL,
2006).

6 See Sam Hind, 'Digital Navigation and the Driving-Machine:
Supervision, Calculation, Optimization, and Recognition', *Mobilities*,
xiv/4 (2019), pp. 401–17.

7 As I have shown in a different navigational context, that of navigating
inland waterways, the materiality of the spiral-bound map book is
integral to the way it is used in practice. Mike Duggan, 'Cruising
Landscape-Objects: Inland Waterway Guidebooks and Wayfinding
with Them', *Cultural Geographies*, xxix/2 (2021), pp. 167–83.

8 Relatedly, black-cab drivers training for the Knowledge are also
required to develop a shorthand vocabulary for calling out routes
during their exams. See Lordan, *The Knowledge*.

9 As Hind notes in 'Digital Navigation and the Driving-Machine', p. 403, 'mapping – and more precisely, vehicular navigation – has always been a hybrid endeavour.' See also Tim Dant, who argues that in the practice of driving, neither car nor driver can be considered separate entities; they are, instead, always-already driver-car. Dant, 'The Driver-Car', *Theory, Culture and Society*, XXI/4–5 (2004), pp. 61–79.

10 For an introduction and background to rally-car racing, see Hans Erik Næss, *A Sociology of the World Rally Championship* (London, 2014). For an introduction to co-driver navigation, see Martin Holmes, *Rally Navigation: Develop Winning Skills with Advice from the Experts* (Yeovil, 1997).

11 Rally races often take place on public roads, and it is only during the race itself that the drivers can disregard the speed limits of the road.

12 Many examples of the driver/co-driver partnership in action are available to view online.

13 See Pavel Dresler, 'Secrets of a Rally Navigator', Škoda Motorsport Storyboard, www.skoda-storyboard.com, 28 October 2018.

14 See, for example, Jan Dirk Kruit et al., 'Design of a Rally Driver Support System Using Ecological Interface Design Principles', IEEE *Transactions on Systems, Man, and Cybernetics, Part B (Cybernetics)*, XXXV/1 (February 2005), pp. 1235–9. The Jemba Inertia Notes System has been popular among amateur rallying communities in North America, Europe and New Zealand. Co-driver training has also turned to digital simulator technology in recent years.

15 There is an interesting absence of companies attempting to use Lidar and machine-learning mapping technology in this form of driving, considering how prominent the technology has become for other driving practices (see Chapter Six).

16 Joseph J. Amato, *On Foot: A History of Walking* (New York, 2004); Steve Boga, *Orienteering* (Mechanicsburg, PA, 1997).

17 Jennie Middleton's recent account of walking as a deeply social and cultural activity encapsulates some of this complexity: Jennie Middleton, *The Walkable City: Dimensions of Walking and Overlapping Walks of Life* (Abingdon, 2022).

18 There is very little research into navigation and those unable to walk, especially within the social sciences. I can, however, recommend the Global Disability Innovation Hub (www.disabilityinnovation.com) for insight into what is being done to improve the navigational experiences of differently abled people.

19 Many of these earlier forms were developed and became popular alongside the rise in urban tourism in the twentieth century. See

Mark Monmonier, ed., *The History of Cartography*, vol. VI: *Cartography in the Twentieth Century* (Chicago, IL, 2015).

20 For an overview, see Greg Milner, *Pinpoint: How GPS Is Changing Technology, Culture, and Our Minds* (London, 2017); and Stuart Dunn, *A History of Place in the Digital Age* (Abingdon, 2019).

21 Mike Duggan, 'Mapping Interfaces: An Ethnography of Everyday Digital Mapping Practices', PhD thesis, Royal Holloway, University of London, 2017.

22 This idea was popularized by anthropologists studying the Indigenous practices of wayfinding around the world, and it has since come to be used to understand navigational practices in a broader sense. While anthropological studies have been useful in developing knowledge of wayfinding, they have also been heavily criticized for co-opting for academic gain what has long been known and practised by Indigenous peoples. As discussed in the Introduction, non-Indigenous peoples have no automatic right to discuss and disseminate Indigenous wayfinding knowledge, because it is often regarded as deeply held cultural knowledge.

23 Eric Laurier, Barry Brown and Moira McGregor, 'Mediated Pedestrian Mobility: Walking and the Map App', *Mobilities*, XI/1 (2016), pp. 117–34.

24 Clancy Wilmott, *Mobile Mapping: Space, Cartography and the Digital* (Amsterdam, 2020).

25 Matthew Zook and Mark Graham, 'Mapping DigiPlace: Geocoded Internet Data and the Representation of Place', *Environment and Planning B: Planning and Design*, XXXIV/3 (2007), pp. 466–82; Matthew Zook and Mark Graham, 'The Creative Reconstruction of the Internet: Google and the Privatization of Cyberspace and DigiPlace', *Geoforum*, XXXVIII/6 (2007), pp. 1322–43.

26 See also those working to 'decolonize' the map by removing and replacing colonial place names with their Indigenous origins. For example, Ruben Rose-Redwood et al., 'Decolonizing the Map: Recentering Indigenous Mappings', *Cartographica*, LV/3 (2020), pp. 151–62.

27 Lucas Chancel, *Unsustainable Inequalities: Social Justice and the Environment* (Cambridge, MA, 2020).

28 Collecting, analysing and packaging data in order to generate economic value from the prediction of future trends in society is a key business model for location-based services, such as navigational apps. We see this extraction > value-generation process across the data economy, and it would not at all be a surprise to learn that navigation

apps were sharing (or selling) packaged data trends about the urban environment to urban planners and the like.

29 Many of these apps are designed with women walking alone in mind, although that simplifies the range of users and uses.

30 See www.police.uk, accessed July 2022.

31 Kaushiki Das, 'Digital "Lakshman Rekhas": Understanding the Impact of Safety Apps on Women's Freedom of Movement in Urban Spaces', *Global Perspectives*, 11/1 (2021), article no. 25608.

32 Ibid.

33 With this in mind, feminist activist groups have developed alternative safety apps that attempt to take these wider factors into account. See, for example, Free to Be, a crowd-mapping tool that was developed in collaboration with CrowdSpot, Monash University's XYX Lab.

34 Known as the 'labelling theory of crime'. A great deal has been written on the topic in the fields of sociology and criminology, starting most notably with Howard Becker and Erving Goffman in the 1960s and '70s. There have since been many iterations and divergences of the theory in these fields, but the central idea still holds currency in the way we might understand how and why certain places first become associated with and then maintain their association with crime. Despite the prevalence of these ideas, it must, however, be noted that they originate and circulate primarily in the discourse on cities in the United States and Europe, and therefore do not reflect a global understanding.

35 See, for example, Svati Kirsten Narula, 'The Real Problem with a Service Called "Ghetto Tracker"', *The Atlantic*, www.theatlantic.com, 6 September 2013; and Agnieszka Leszczynski, 'Speculative Futures: Cities, Data, and Governance beyond Smart Urbanism', *Environment and Planning A: Economy and Space*, XLVIII/9 (2016), pp. 1691–708. Leszczynski also discusses the link between safety apps, data, geo-surveillance and the way these can produce speculative urban futures.

36 L. Maxwell et al., 'A Content Analysis of Personal Safety Apps: Are They Keeping Us Safe or Making Us More Vulnerable?', *Violence against Women*, XXVI/2 (2020), pp. 233–48.

2 Interfaces of Movement

1 For much more on the topic, see Tamara Chin, 'The Invention of the Silk Road, 1877', *Critical Inquiry*, XL/1 (2013), pp. 194–219.

2 Manifest, for example, is a supply-chain mapping platform built by researchers from New York University. It is probably the most up-to-date map showing the movements of global supply chains.

3 Much of this tracking is done in-house by multinational logistics companies, such as Maersk, Evergreen and MSC, and built by such companies as IBM, SAP and Oracle, but there are also independent companies competing for a piece of the pie. See, for example, the Live Earth platform, www.liveearth.com, accessed May 2022.

4 For more on this story, see reporting by Annie Palmer, such as in 'Amazon Uses an App Called Mentor to Track and Discipline Delivery Drivers', CNBC, www.cnbc.com, 12 February 2021.

5 For more on the concept of (im)mobility, see Mimi Sheller, *Mobility Justice: The Politics of Movement in an Age of Extremes* (New York, 2018).

6 For much more on the topic, see Laura Vaughan, *Mapping Society: The Spatial Dimensions of Social Cartography* (London, 2018).

7 See contributions to Laura Bliss, *The Quarantine Atlas: Mapping Global Life under COVID-19* (New York, 2022).

8 See Kimbal Quist Bumstead and Sol Perez-Martinez, 'Dreaming of a Post-COVID World', www.livingmaps.org, accessed 3 March 2023; and the Corona Diaries mapping project, www.coronadiaries.io, accessed January 2022.

9 In reality, this would have been very unlikely. Once GPS had been made available to commercial organizations, it entered society in such a way that it would not easily be retracted without serious political intervention. Since the mid-1990s it has never been 'switched off'. To do so today would cause chaos owing to the number of applications that use it.

10 Missiles guided by GPS set a new tone for warfare during the Gulf War. Today GPS is used extensively by military organizations across the world, in weapons technology, surveillance and supply chains.

11 Nitin Govil, 'Something Spatial in the Air: In-Flight Entertainment and the Topographies of Modern Air Travel', in *MediaSpace: Place, Scale and Culture in a Media Age*, ed. Nick Couldry and Anna McCarthy (London, 2004), pp. 233–54.

12 Many societies around the world have become used to the idea that quantifiable observations offer us an objective view of ourselves, which we can then use to improve ourselves. This is not just applied to health and fitness; it has become one of the prevalent logics of science and society in recent decades. There is a large body of research that has unpacked the history of this and the way technology has shaped its development. Deborah Lupton, *The Quantified Self* (Cambridge, 2016) and Btihaj Ajana, ed., *Self-Tracking: Empirical and Philosophical Investigations* (London, 2018) are useful places to start.

13 See, for example, the work of Wally GPX and Anthony Hoyte.

14 For more on this story, see Alex Hern, 'Fitness Tracking App Strava Gives Away Location of Secret u.s. Army Bases', *The Guardian*, www.theguardian.com, 28 January 2018.

15 For more on this story, see the investigative reporting by Johana Bhuiyan, 'The New Warrant: How u.s. Police Mine Google for Your Location and Search History', *The Guardian*, www.theguardian.com, 16 September 2021.

16 See James Cheshire and Oliver Uberti, *Where Animals Go: Tracking Wildlife with Technology* (London, 2016).

17 See, for example, the bold claims made by the ICARUS initiative.

18 See Artemis Skarlatidou and Muki Haklay, *Geographic Citizen Science Design: No One Left Behind* (London, 2021).

19 See, for example, *The Secret Life of the Cat* (BBC, 2012–13).

20 Jennifer Gabrys, *Program Earth* (Boston, MA, 2016).

21 Data visualization, or dataviz, is a burgeoning field in itself, and many people now use innovative digital and non-digital methods for telling stories with data. This field includes mapping, but there are many other ways in which 'data stories' are being told. See the work of the Public Data Lab (www.publicdatalab.org), and Helen Kennedy and Martin Engebretsen, *Data Visualization and Society* (Amsterdam, 2020).

22 Alison Powell, *Undoing Optimization: Civic Action in Smart Cities* (New Haven, CT, and London, 2021), p. 142.

23 As the climate crisis has emerged into public consciousness, so has a lay and intellectual desire to 'get back to Earth'. This can be seen in theoretical discussions about the need to return to Earth (see Bruno Latour, *Down to Earth: Politics in the New Climatic Regime* (Cambridge, 2018)) and in the calls to rewild the countryside and develop organic farming methods. This has mostly been a preoccupation of Western cultures, which have noticed how capitalist neglect for the Earth has led to a detachment from the land that supports life. This discourse – often purported to be a revolution – has long been the bedrock of many Indigenous peoples' relationship to the Earth.

24 There have been considerable efforts to 'modernize' farming through mapping technology over the past century. Throughout the 2010s this has been exacerbated by drone and AI mapping technology. For an up-to-date overview see Panel for the Future of Science and Technology, *Artificial Intelligence in the Agri-Food Sector: Applications, Risks and Impacts* (European Parliamentary Research Service, 2023).

25 See Alisa Reznick, 'Border Wall Scars: "It Feels Like If Someone Got a Knife and Dragged It across My Heart"', *High Country News*, 4 February 2021.

26 Ibid.
27 Chelsea Steinauer-Scudder, 'Counter Mapping', *Emergence*, https://emergencemagazine.org, 8 February 2018.
28 See Candace Fujikane, *Mapping Abundance for a Planetary Future* (Durham, NC, 2021).
29 For much more on the topic of digital bordering, see Louise Amoore, *The Politics of Possibility: Risk and Security beyond Probability* (Durham, NC, 2013); and Lilie Chouliaraki and Myria Georgiou, 'The Digital Border: Mobility beyond Territorial and Symbolic Divides', *European Journal of Communication*, XXXIV/6 (2019), pp. 594–605.
30 See Associated Press, 'Hungarian Police Find Two Tunnels Used by Migrants on Border', *Associated Press*, www.apnews.com, 29 November 2019.

3 Mapping Power and Politics

1 Steve Bickerstaff, *Election Systems and Gerrymandering Worldwide* (New York, 2020); Stan Hok-Wui, 'Gerrymandering in Electoral Autocracies: Evidence from Hong Kong', *British Journal of Political Science*, XLIX/2 (2017), pp. 579–610.
2 See the work of Brian Harley, Denis Wood and Mark Monmonier for some of the best-known 'critical cartographers' to work on this topic in the 1980s and '90s. Brian Harley, 'Deconstructing the Map', *Cartographica*, XXVI/2 (1989), pp. 1–20; Denis Wood, *The Power of Maps* (London, 1992); Mark Monmonier, *How to Lie with Maps* (Chicago, IL, 1996). However, as the chronicler of critical cartography Jeremy Crampton has argued, some of these ideas go back as far as the 1940s: Jeremy Crampton, *Mapping: A Critical Introduction to Cartography and GIS* (Chichester, 2010). Overlapping with this work at the same time were feminist geographers and philosophers studying how maps were produced as a form of situated knowledge and power-representation. All too often these voices are forgotten in the history of 'critical cartography'. See Donna Haraway, 'Situated Knowledges: The Science Question in Feminism and the Privilege of Partial Perspective', *Feminist Studies*, XIV/3 (1988), pp. 575–99; Doreen Massey, *Space, Place and Gender* (Minneapolis, MN, 1994).
3 The philosopher Bruno Latour is regarded by many as one of the first to recognize formally how the power of maps was produced, distributed and circulated in society. See Bruno Latour, *Science in Action* (Cambridge, MA, 1987). Nevertheless, we should also assume that this had been recognized by others outside the academy long before then.

4 See, for example, the BBC's Mapping the World collection of programmes; *The West Wing* episode 'Somebody's Going to Emergency, Somebody's Going to Jail' (2001); Tim Marshal, *The Power of Geography: Ten Maps that Reveal the Future of our World* (London, 2021).

5 Recognized by Brian Harley as 'internal' and 'external' sources of power: Harley, 'Deconstructing the Map', pp. 1–20.

6 There is a whole genre of work that examines how different map projections offer a different perspective on the world, including the size and dimensions of countries, distances between points and their position on the map. Much of this is down to the difficulty of representing the sphere of the Earth on a flat surface. It is widely covered elsewhere and does not need to be repeated here.

7 In recent iterations of Ordnance Survey maps, crosses have, in *some* cases, been replaced by 'PW' (place of worship), which is a step in the right direction.

8 See Haraway, 'Situated Knowledges'.

9 See Jorge Luis Borges, 'On Exactitude in Science' [1946], in *Collected Fictions*, trans. A. Hurley (London, 1999), pp. 325–6.

10 See, for example, Juan Felipe Forero Duarte, 'Composition Is Also Coordination: Multi-Actor Digital Mapping in Madrid', *Livingmaps Review*, XI (2021).

11 Mike Duggan, 'Cultures of Enthusiasm: An Ethnographic Study of Amateur Map Maker Communities', *Cartographica*, LIV/3 (2019), pp. 217–29.

12 Jeremy Crampton was one of the first to bring this to our attention. See his 'Maps as Social Constructions: Power, Communication and Visualization', *Progress in Human Geography*, XXV/2 (2001), pp. 235–52.

13 Chris Perkins, 'Cultures of Map Use', *Cartographic Journal*, XLV/2 (2008), pp. 150–58.

14 Google Maps is not just one map, but rather a series of stitched-together images forever in flux and updated in near real time. For much more on this and its origins, see Laura Kurgan, *Close Up at a Distance: Mapping, Technology, Politics* (New York, 2013).

15 It is well known that one of the ways Google maintains its dominance in this market is by buying up competitors and start-ups experimenting with new mapping technology. There is even a semi-fictional Netflix series (*The Billion Dollar Code*, 2021) on some of the more underhand tactics the company allegedly used to maintain its market share.

16 Branded pins come and go on Google Maps. Like many other features on these maps, they are rarely permanent and always

subjected to tweaks or removal. It is never entirely clear why, but one can assume it is because some features are used more than others, or that some generate more revenue than others.

17 This is done primarily through the Google AdWords and AdSense programs.

18 Google Maps and the data it collects must always be considered in relation to the other Google products and services that are available. The data Google collects about its users is not left in product or service silos, but rather is pooled and analysed together in relational ways, so that what we search for in Google Maps has a connection to what we search for with Google and on YouTube, or what we say in Gmail messages.

19 See, for example, Hexagon Geospatial (www.hexagon.com), which specializes in selling this idea to governments and organizations. This follows a deal made in 2012, in which Google allowed the World Bank to give away the platform's source data and Map Maker API to countries for the purposes of economic development and managing humanitarian disasters.

20 See Craig Dalton, 'For Fun and Profit: The Limits and Possibilities of Google Maps-Based Geoweb Applications', *Environment and Planning A*, XLVII (2015), pp. 1029–46; Scott McQuire, 'One Map to Rule Them All? Google Maps as Digital Technical Object', *Communication and the Public*, IV/2 (2019), pp. 150–65.

21 See the Native Land project (www.native-land.ca) for an overview of where the lands of Indigenous peoples are in the United States and around the world. This map is striking in many ways, not least for challenging the common assumption that all territory must be clearly marked and bounded. See also Corrina Gould's essay 'Ohlone Geographies' in the Anti-Eviction Mapping Project, *Counterpoints: A San Francisco Bay Area Atlas of Displacement and Resistance* (Oakland, CA, 2021), for a history of how the lands of the Ohlone people were stolen and colonized. For other decolonial mapping projects, please see This is Not an Atlas (www.notanatlas.org) and The Decolonial Atlas (www.decolonialatlas.com)

22 Valentina Carraro, 'A Glitch in Google Maps', in *Jerusalem Online: Critical Cartography for the Digital Age*, ed. Valentina Carraro (London, 2021).

23 They are, of course, not the same maps at all, although they share the same cartographic style. Every time Google Maps is launched, the viewer may see a slightly different map from the time before, owing to the many changes that will have been made since then. It has been

reported that Google makes millions of changes to its maps each day. See Ethan Russell (Google Maps Product Director), '9 Things to Know about Google's Maps Data: Beyond the Map', *Google Cloud Blog*, https://cloud.google.com/blog, 1 October 2019.

24 See Mateusz Fafinski, 'In Putin's War, the Map Is Not the Territory', *Foreign Policy*, www.foreignpolicy.com, 7 March 2022.

25 See Andrés Luque-Ayala and Flávia Neves Maia, 'Digital Territories: Google Maps as a Political Technique in the Re-Making of Urban Informality', *Environment and Planning D: Society and Space*, XXXVII/3 (2019), pp. 449–67.

26 Ibid., p. 463.

27 There is a debate to be had here about whether this constitutes an exchange of labour for the service. Many would argue that users of platforms are actually workers not receiving adequate pay. See, for example, Christian Fuchs, *Digital Capitalism: Media, Communication and Society* (Abingdon, 2022).

28 See Monmonier, *How to Lie with Maps*.

29 See Claire Reddleman, *Cartographic Abstraction in Contemporary Art: Seeing with Maps* (Abingdon, 2019); Henk van Houtum and Rodrigo Bueno Lacy, 'The Migration Map Trap: On the Invasion Arrows in the Cartography of Migration', *Mobilities*, XV/2 (2020), pp. 196–219; Maite Vermeulen, Leon de Korte and Henk van Houtum, 'How Maps in the Media Make Us More Negative about Migrants', *The Correspondent*, www.thecorrespondent.com, 2 September 2020.

30 See Alexander Kent, 'Political Cartography: From Bertin to Brexit', *Cartographic Journal*, LIII/3 (2016), pp. 199–201.

31 See Guntram Henrik Herb, *Under the Map of Germany: Nationalism and Propaganda, 1918–1945* (London, 1997).

32 See the collection of images, which includes many maps, submitted to the UK government's Department for Digital, Culture, Media and Sport's (DCMS) enquiry into fake news following this scandal: DCMS, 'Disinformation and "fake news": Final Report', *House of Commons*, HC 1791 (2019), https://publications.parliament.uk/pa/cm201719/cmselect/cmcumeds/1791/1791.pdf.

33 Laura Lo Presti, 'The Migrancies of Maps: Complicating the Critical Cartography and Migration Nexus in "Migro-Mobility" Thinking', *Mobilities*, XV/6 (2020), pp. 911–29.

34 See, for example, Daniel McLaughlin, 'Mass Migration Guided by Mobiles and Social Media', *Irish Times*, 9 September 2015; BBC, 'Migrant Crisis: "We Would Be Lost without Google Maps"', www.youtube.com, 9 September 2015.

35 Migrants who use digital technology to help them have been labelled 'digital refugees'. In popular Western news media, reporters often express surprise that migrants use such technology as the smartphone, maps and social media to assist them in communicating and planning their journeys. This shows a lack of knowledge about who uses digital technology today, and perpetuates the media trope that migrant refugees come from dated societies. See Hannah A. Gough and Katherine V. Gough, 'Disrupted Becomings: The Role of Smartphones in Syrian Refugees' Physical and Existential Journeys', *Geoforum*, CV (2019), pp. 89–98; Heike Graf, 'Media Practices and Forced Migration: Trust Online and Offline', *Media and Communication*, VI/2 (2018), pp. 149–57; and Tim Adams, 'Inside Dunkirk's Desperate Refugee Camps: "They Take Risks because They Feel They Have No Choice"', *The Guardian*, 27 November 2021.

36 See, for example, the Anti-Eviction Mapping Project, *Counterpoints*; Bjørn Sletto et al., *Radical Cartographies: Participatory Map Making from Latin America* (Austin, TX, 2020); Phil Cohen and Mike Duggan, ed., *New Directions in Radical Cartography: Why the Map Is Never the Territory* (London, 2021).

37 There is a lot of research into this topic under the banner of 'critical border studies'. Two notable examples are Louise Amoore, 'Biometric Borders: Governing Mobilities in the War on Terror', *Political Geography*, XXV (2006), pp. 336–51; and Burcu Toğral Koca, 'Bordering Processes through the Use of Technology: The Turkish Case', *Journal of Ethnic and Migration Studies*, XLVIII (2020), pp. 1909–26.

4 Mapping Culture

1 For a more comprehensive history, see Pascal Pannetier and Etienne Houdoy, 'Michelin Guides', in *The History of Cartography*, vol. VI: *Cartography in the Twentieth Century*, ed. Mark Monmonier (Chicago, IL, 2015), pp. 878–83.

2 Although the Ordnance Survey has been making maps since 1791, it was not until the twentieth century that it got a foothold in the mass market, and thereafter in the popular imagination.

3 A large part of the OS business model today is the sale and analysis of geospatial data to commercial buyers. The production of paper maps, although still an important part of the brand, is no longer its core focus. This is owing to the competitive market in everyday digital and mobile maps (now dominated by technology platforms), the growth in the value of geospatial data, and the OS's transition to privately held state-owned company, starting in 2015.

4 Mark Graham and Martin Dittus, *Geographies of Digital Exclusion: Data and Inequality* (London, 2022).

5 For more on cartographic abstractions, see Claire Reddleman, *Cartographic Abstraction in Contemporary Art: Seeing with Maps* (Abingdon, 2018); Gunnar Olsson, *Abysmal: A Critique of Cartographic Reason* (Chicago, IL, 2010); and Laura Kurgan, *Close Up at a Distance: Mapping, Technology, and Politics* (New York, 2013).

6 Mike Duggan, 'The Cultural Life of Maps: Everyday Place-Making Mapping Practices', in *New Directions in Radical Cartography: Why the Map Is Never the Territory*, ed. Phil Cohen and Mike Duggan (London, 2021), pp. 3–20.

7 See Marie Patino, 'MapLab: Thirty Days of Maps', *Bloomberg*, 29 December 2021.

8 For a history of women in Western cartography, see Alice Hudson and Mary Ritzlin, 'Preliminary Checklist of Pre-Twentieth-Century Women in Cartography', *Cartographica*, XXXVII/3 (2000), pp. 3–8; Christina E. Dando, *Women and Cartography in the Progressive Era* (Abingdon, 2018); Judith Tyner, *Women in American Cartography: An Invisible Social History* (London, 2019); and Will C. van den Hoonaard, *Map Worlds: A History of Women in Cartography* (Waterloo, ON, 2013).

9 In common with most professions, cartography has many problems caused by gender inequality. These are being addressed as they are discussed and acted on in workplaces, community forums, academic debates and industry magazines, and on social media. Nevertheless, there is much work to be done before it can be said that cartography has gender parity. See, for example, the International Cartographic Association's commission on Gender and Cartography (2003–7) and the work of the Society for Women Geographers (www.iswg.org).

10 Doreen Massey, *Space, Place and Gender* (Minneapolis, MN, 1994). See also the work of the geographer Gillian Rose, especially her *Visual Methodologies* (London, 2016).

11 For a detailed discussion of how gender is (re)produced through digital mapping technology, or what is sometimes called 'new spatial media', see Agnieszka Leszczynski and Sarah Elwood, 'Feminist Geographies of New Spatial Media', *Canadian Geographer*, LIX/1 (2015), pp. 12–28.

12 This is not to say that men do not navigate with safety in mind, nor that 'men' or 'women' are homogeneous in their map-reading strategies, but rather to suggest that the safety concerns of men and women when reading maps to navigate the city are usually different, based on their life experiences.

13 Alexander P. Boone, Xinyi Gong and Mary Hegarty, 'Sex Differences in Navigation Strategy and Efficiency', *Memory and Cognition*, XLVI (2018), pp. 909–22.

14 For an introduction to this field, see Marianna Pavlovskaya, 'Feminism, Maps and GIS', in *International Encyclopedia of Human Geography*, ed. Rob Kitchin and Nigel Thrift (London, 2009), pp. 37–43.

15 Rebecca Solnit, *Infinite City: A San Francisco Atlas* (Los Angeles, CA, 2010).

16 Ibid., p. 48.

17 For more details, see Immerse News, 'Co-Creating a Map of Queer Experience: An Interview with Lucas LaRochelle', *Immerse News*, www.immerse.news, 2 November 2019.

18 There is a correlation between the rise in land values and the rise in a booming tech culture that has come to characterize the city since the turn of the millennium. See Sharon Zukin, *The Innovation Complex: Cities, Tech, and the New Economy* (New York, 2020).

19 Counter-mapping has a number of monikers, including critical cartography and critical GIS, participatory mapping and public geography.

20 Full list of contributors: Dr Pat Noxolo, Dr Tia Monique Uzor, Obafeyikemi Luther, Evie Otoki, Lea Bematol and Bea Tizzy, with contributions by the staff of the RGS: Dr Catherine Souch and Sarah Evans. Gennaro Ambrosino has also produced a film of the installation, available at www.rgs.org/research/higher-education-resources/dreading-the-map, accessed January 2022.

21 See Phil Cohen, 'Dread Maps: Interview with Sonia E. Barrett', *Livingmaps Review*, 11 (2021).

22 Chris Perkins, 'Map Collecting Practices', in *Advances in Cartography and GIscience*, ed. Anne Ruas, 2 vols (Berlin, 2011, vol. 1), pp. 133–46.

23 For a detailed account and history of map societies, see Robert W. Karrow, 'Societies, Map', in *Cartography in the Twentieth Century*, ed. Monmonier, pp. 1419–21.

24 Although, of course, this does not mean that all norms and values are agreed upon or respected equally by the community. Cultures are not homogeneous groups, despite the inclination to treat them as such. They share some commonalities in a world of difference, and should therefore not be considered unified on all subjects.

25 A large body of research was conducted after the initial explosion of Pokémon Go, covering its impact on the urban environment, its role in the growing popularity of augmented reality (AR) technology, and how particular forms of socializing and communication emerged from

its use. For an introduction to this work, see Larissa Hjorth and Ingrid Richardson, 'Pokémon Go: Playful Phoneurs and the Politics of Digital Wayfarers', *Mobile Media and Communication*, v/1 (2017), pp. 3–14.

26 This was not the last time Niantic was sued. There are numerous cases still live relating to users attempting to reach Pokémon on the map by trespassing on private lands, and by resident groups seeking to get virtual Pokémon removed from their street because of the unwanted visitors they are perceived to attract.

27 The impact of humanitarian maps produced by volunteers from a distance is a hotly debated topic. Many people, including charity and humanitarian organizations, place great value on them, arguing that they offer a quick and efficient way to make and distribute maps to people on the ground. Others have questioned how this type of aid perpetuates the status quo of aid-giving from afar, and how crowdsourcing can lead to significant cartographic errors.

28 For an academic analysis of my research here, see Mike Duggan, 'Cultures of Enthusiasm: An Ethnographic Study of Amateur Map-Maker Communities', *Cartographica*, LIV/3 (2019), pp. 217–29.

29 Doreen Massey, *For Space* (London, 2005).

5 Maps that Make the Money Go Round

1 Sybil E. Crowe, *The Berlin West African Conference, 1884–1885* (London, New York and Toronto, 1942); Thomas Pakenham, *The Scramble for Africa: 1876–1912* (London, 2003).

2 James C. Scott, *Seeing Like a State: How Certain Schemes to Improve the Human Condition Have Failed* (New Haven, CT, 1999).

3 Candace Fujikane, *Mapping Abundance for a Planetary Future* (Durham, NC, 2021).

4 Irendra Radjawali and Oliver P. June, 'Counter-Mapping Land Grabs with Community Drones in Indonesia', *Land Grabbing, Conflict and Agrarian-Environmental Transformations: Perspectives from East and Southeast Asia Conference*, Conference Paper no. 80 (5–6 June 2015).

5 It could be argued that this was political posturing from the start, because for many Indigenous communities cartographic maps could never represent the complex relationships they have with the land. For an overview of the policy, see Nabiha Shahab, 'Indonesia: One Map Policy', Open Government Partnership, December 2016, available at www.opengovpartnership.org.

6 Ibid., p. 5.

7 In a separate mapping project, the government has since published its first map of customary forests, which has given some hope to

those who want Indigenous land rights to be included on official maps. However, at the time of writing debate is ongoing about the effectiveness of this approach in protecting these lands from economic development activities, particularly as the bureaucratic process of getting land recognized in the first place is fraught with difficulty.

8 Aryo Adhi Condro et al., 'Retrieving the National Main Commodity Maps in Indonesia Based on High-Resolution Remotely Sensed Data Using Cloud Computing Platform', *Land*, IX/10 (2020), p. 377.

9 The trend for Indigenous counter-mapping has emerged disproportionately all over the world since the 1960s. For an overview, see Mac Chapin, Zachary Lamb and Bill Threlkeld, 'Mapping Indigenous Lands', *Annual Review of Anthropology*, XXXIV (2005), pp. 619–38.

10 Irendra Radjawali, Oliver Pye and Michael Flitner, 'Recognition through Reconnaissance? Using Drones for Counter-Mapping in Indonesia', *Journal of Peasant Studies*, XLIV/4 (2017), pp. 817–33. For an insight into others using drones in Indigenous mapping contexts, see the work of Jaime Paneque-Gálvez, who has worked extensively with communities in South America. An audio interview about his work is available here: Jaime Paneque-Gálvez, Trishant Simlai and Jennifer Gabrys, 'Jaime Paneque-Gálvez: Community Forest Monitoring with Drones', Smart Forests Atlas, https://atlas.smartforests.net/en/radio/jaime-paneque-galvez, 10 August 2022.

11 Nancy Lee Peluso, 'Whose Woods Are These? Counter-Mapping Forest Territories in Kalimantan, Indonesia', *Antipode*, XXVII/4 (1995), pp. 383–406.

12 Jefferson Fox et al., 'Mapping Boundaries, Shifting Power: The Socio-Ethical Dimensions of Participatory Mapping', in *Contentious Geographies: Environmental Knowledge, Meaning, Scale*, ed. Michael K. Goodwin, Maxwell T. Boykoff and Kyle T. Evered (Aldershot, 2008), pp. 203–17.

13 See Pip Thornton, 'A Critique of Linguistic Capitalism: Provocation/Intervention', *Geohumanities*, IV/2 (2018), pp. 417–37.

14 See Rowman Wilken, *Cultural Economies of Locative Media* (Oxford, 2020); and Harrison Smith, 'Metrics, Locations, and Lift: Mobile Location Analytics and the Production of Second-Order Geodemographics', *Information, Communication and Society*, XXII/8 (2017), pp. 1044–61.

15 There are now many books on the impact of machine learning on society. For one of the best critical discussions, see Louise Amoore, *Cloud Ethics: Algorithms and the Attributes of Ourselves and Others* (Durham, NC, 2020).

16 The phrase has since been used by countless others. It is also debated heavily; many argue that it is a phrase without much meaning, especially when we consider that data is not a finite resource as oil is, that it is easily transferable and replicable, and that its value is not determined at the level of global markets, but instead agreed upon depending on its type, size and potential use.

17 UK Government, 'Unlocking the Power of Location: The UK's Geospatial Strategy, 2020 to 2025' (Cabinet Office, 2020), p. 5, available at www.gov.uk.

18 Fortune Business Insights, 'Locations Analytics Market Research Report' (2022), available at www.fortunebusinessinsights.com/information-and-technology-industry.

19 Pooja Iyer et al., 'Location-Based Targeting: History, Usage, and Related Concerns', Center for Media Engagement, University of Texas at Austin (2021), available at www.mediaengagement.org/research.

20 For an introduction to the world of the data imaginaries into which this industry buys, see David Beer, *The Data Gaze* (London, 2019).

21 For an introduction to the ways that data shapes everyday practices and how everyday practices shape the collection of data, see Helen Kennedy, 'Living with Data: Aligning Data Studies and Data Activism through a Focus on Everyday Experiences of Datafication', *Krisis: Journal for Contemporary Philosophy*, 1 (2018), pp. 18–30; and Noortje Marres, *Digital Sociology: The Reinvention of Social Research* (London, 2017).

22 Rob Kitchin, *The Data Revolution: A Critical Analysis of Big Data, Open Data and Data Infrastructures* (London, 2021).

23 Claudia Aradau and Tobias Blanke, *Algorithmic Reason: The New Government of Self and Other* (Oxford, 2022).

24 Forbrukerrådet, 'Out of Control: How Consumers Are Exploited by the Online Advertising Industry', *Forbrukerrådet* (2020), available at https://fil.forbrukerradet.no.

25 Byron Tau and Georgia Wells, 'Grindr User Data Was Sold through Ad Networks', *Wall Street Journal*, www.wsj.com, 2 May 2022.

26 The Pillar, 'Pillar Investigates: USCCB Gen Sec Burrill Resigns after Sexual Misconduct Allegations', *The Pillar*, www.pillarcatholic.com, 20 July 2021.

27 Jennifer Valentino-DeVries et al., 'Your Apps Know Where You Were Last Night, and They're Not Keeping It Secret', *New York Times*, 10 December 2018. See also the ongoing *New York Times* series 'The Privacy Project'.

28 John McDaniel and Ken Pease, *Predictive Policing and Artificial Intelligence* (Abingdon, 2021); Sarah Lamdan, 'When Westlaw Fuels

ICE Surveillance: Legal Ethics in the Era of Big Data Policing',
NYU Review of Law and Social Change, XLIII/2 (2019), pp. 255–93.

29 Privacy International, 'Digital Stop and Search: How the UK Police
Can Secretly Download Everything from Your Mobile Phone',
Privacy International, https://privacyinternational.org, 27 March 2018.

30 Jenna McLaughlin, 'How Chicago Police Convinced Courts
to Let Them Track Cellphones without a Warrant', *The Intercept*,
www.theintercept.com, 18 October 2016.

31 See www.predpol.com/about, accessed July 2022.

32 Simon Egbert and Matthias Leese, *Criminal Futures: Predictive
Policing and Everyday Police Work* (Abingdon, 2021).

33 See, for example, Rashida Richardson, Jason M. Schultz and Kate
Crawford, 'Dirty Data, Bad Predictions: How Civil Rights Violations
Impact Police Data, Predictive Policing Systems, and Justice',
New York University Law Review, 192 (2019), n.p.

34 Ruha Benjamin, *Race after Technology: Abolitionist Tools for the New
Jim Crow Code* (New York, 2019).

35 Amoore, *Cloud Ethics*.

6 Mapping Presents and Futures

1 Martin Placek, 'Size of the global autonomous vehicle market in 2021
and 2022, with a forecast through 2030', *Statista*, www.statista.com,
16 January 2023.

2 Machine learning is the dominant form of artificial intelligence (AI)
that exists today. It makes up what is sometimes called 'Narrow AI',
because it demonstrates the intelligence of computers when applied
to specific cases. 'General AI' is the far broader quest to develop
computers that will one day surpass the intelligence of humans in all
fields. I prefer to use the term 'machine learning' over 'AI', because it
better describes what computers are doing with data today.

3 See Claudia Aradau and Tobias Blanke, *Algorithmic Reason: The New
Government of Self and Other* (Oxford, 2022).

4 Self-driving cars are a competitive market in which each
manufacturer and software developer takes a different approach.
The most notable difference among those aiming for full automation
is between those using Lidar (Light Detection and Ranging)
imaging technology with what is called 'HD [High Definition]
maps', and those primarily processing image data from cameras
using machine-learning technology. For an overview of how the
different technical systems work, see V. Shreyas et al., 'Self-Driving
Cars: An Overview of Various Autonomous Driving Systems', in

Advances in Data and Information Sciences, ed. Shailesh Tiwari et al., vol. XCIV (2020).

5 Sam Hind, 'Machinic Sensemaking in the Streets: More-than-Lidar in Autonomous Vehicles', in *Seeing the City Digitally: Processing Urban Space and Time*, ed. Gillian Rose (Amsterdam, 2022), pp. 57–80.

6 Ibid.

7 Louise Amoore, *Cloud Ethics: Algorithms and the Attributes of Ourselves and Others* (Durham, NC, 2020).

8 See Salman Azhar, 'Building Information Modeling (BIM): Trends, Benefits, Risks, and Challenges for the AEC Industry', *Leadership and Management in Engineering*, XI/3 (2011), pp. 241–52.

9 Monica Degen, Claire Melhuish and Gillian Rose, 'Producing Place Atmospheres Digitally: Architecture, Digital Visualization Practices and the Experience Economy', *Journal of Consumer Culture*, XVII/1 (2017), pp. 3–24; David Bissell and Gillian Fuller, 'Material Politics of Images: Visualising Future Transport Infrastructures', *Environment and Planning A*, XLIX/11 (2007), pp. 2477–96.

10 James Bridle, 'Render Ghosts', *Electronic Voice Phenomena*, 14 November 2013. See also www.render-search.jamesbridle.com, accessed December 2021.

11 See Matthew Mindrup, *The Architectural Model: Histories of the Miniature and the Prototype, the Exemplar and the Muse* (Boston, MA, 2019).

12 The geographical data these companies collect includes local and national map databases, image and location data collected from vehicles and smartphones, proprietary data sets acquired through acquisition and payment, data volunteered by users and data submitted when using other connected services.

13 Google's 'AI for Social Good' project is already doing this elsewhere on the African continent, in Ghana.

14 UNHCR, Operational Data Portal: Uganda (2023), available at https://data.unhcr.org/en/country/uga, accessed July 2022.

15 The availability of satellite imagery has been a game-changer here. The rise of public and private satellites capturing images of the Earth has meant that mapping from afar is a common practice for governments, commercial organizations, not-for-profit organizations and NGOs alike. Rarely is there any discussion about the ethics of doing so. See Laura Kurgan, *Close Up at a Distance: Mapping, Technology, and Politics* (New York, 2013).

16 For more on cartographic abstractions, see ibid.; Claire Reddleman, *Cartographic Abstraction in Contemporary Art: Seeing with Maps*

(Abingdon, 2018); and Gunnar Olsson, *Abysmal: A Critique of Cartographic Reason* (Chicago, IL, 2010).

17 See, for example, Global Partnerships on AI report, 'Climate Change and AI: Recommendations for Government Action', November 2021, available at www.gpai.ai. Curiously, the 2022 Intergovernmental Panel on Climate Change (IPCC) report includes only a brief mention of how AI is being used to map, model and predict the impact of climate change. This is an indication of how climate-science methodology differs significantly from that of other fields of knowledge production, where data science has been adopted to a greater extent.

18 StoryMap is proprietary mapping software made by Esri that allows users to plot stories on the map using interactive technology. Story mapping, however, includes a wider range of mapping software and has a longer history linked to 'deep mapping' digital humanities and spatial humanities projects. See Les Roberts, *Mapping Cultures: Place, Practice, Performance* (London, 2012) for an introduction.

19 In the new millennium there has been a surge in interest in literary geographies, whereby maps have been studied as part of the story as well as being used to map out key events and places in the story. Much of this field has been shaped by digital mapping technology that offers new modes of analysis and representation. For an introduction, see David Cooper, Christopher Donaldson and Patricia Murrieta-Flores, *Literary Mapping in the Digital Age* (Abingdon, 2016). Adjacent to this field is a growing interest in 'comic-book geographies', which use the narrative conventions of the comic strip to map out spatial stories. For an introduction, see Giada Peterle, 'Comics and Maps? A CartoGraphic Essay', *Livingmaps Review*, VII (2019), pp. 1–7.

20 Sébastien Caquard and Stefanie Dimitrovas, 'Story Maps & Co.: The State of the Art of Online Narrative Cartography', *Mappemonde*, CXXI (2017), n.p.

21 Ibid.

22 This is not to say that technology does not mediate and augment these experiences, but rather that there is something specific about cartographic mapping technology and the way it uses abstract representation to distance itself from humans' lived experiences.

23 Chelsea Steinauer-Scudder, 'Counter Mapping', *Emergence*, https://emergencemagazine.org, 8 February 2018.

24 To read more about the differences between the terminology and its uses, see Rob Kitchin, Tracey P. Lauriault and Matthew W. Wilson, ed., *Understanding Spatial Media* (London, 2017).

RESOURCES

Anti-Eviction Mapping Project
www.antievictionmap.com

Bear 71 interactive documentary
https://bear71vr.nfb.ca

CGeomap mapping platform
www.cgeomap.eu

Corona Diaries
www.coronadiaries.io

David Rumsey Map Collection
www.davidrumsey.com

The Decolonial Atlas
www.decolonialatlas.wordpress.com

Humanitarian OpenStreetMap Team
www.hotosm.org

Jeremy Wood artworks
www.gpsdrawing.com

Kate McLean: Sensory Maps
www.sensorymaps.com

Layla Curtis artworks
www.laylacurtis.com

Manifest supply chain mapping project
https://manifest.supplystudies.com

Map-lective
www.map-lective.com

Native Land Digital
www.native-land.ca

OpenStreetMap
www.openstreetmap.org

Polynesian Voyaging Society
www.hokulea.com

Public Data Lab
www.publicdatalab.org

Queering the Map
www.queeringthemap.com

This is Not an Atlas project
www.notanatlas.org

Watch the Med mapping project
www.watchthemed.net

Yassan's GPS drawing project
http://gpsdrawing.info

SELECT BIBLIOGRAPHY

Akerman, James R., *Cartographies of Travel and Navigation* (Chicago, IL, 2006)

Amoore, Louise, *Cloud Ethics: Algorithms and the Attributes of Ourselves and Others* (Durham, NC, 2020)

—, *The Politics of Possibility: Risk and Security beyond Probability* (Durham, NC, 2013)

The Anti-Eviction Mapping Project, *Counterpoints: A San Francisco Bay Area Atlas of Displacement and Resistance* (Oakland, CA, 2021)

Aradau, Claudia, and Tobias Blanke, *Algorithmic Reason: The New Government of Self and Other* (Oxford, 2022)

Beer, David, *The Data Gaze* (London, 2019)

Benjamin, Ruha, *Race after Technology: Abolitionist Tools for the New Jim Crow Code* (New York, 2019)

Bickerstaff, Steve, *Election Systems and Gerrymandering Worldwide* (New York, 2020)

Bliss, Laura, *The Quarantine Atlas: Mapping Global Life under COVID-19* (New York, 2022)

Carraro, Valentina, ed., *Jerusalem Online: Critical Cartography for the Digital Age* (London, 2021)

Cheshire, James, and Oliver Uberti, *Where Animals Go: Tracking Wildlife with Technology* (London, 2016)

Cohen, Phil, and Mike Duggan, eds, *New Directions in Radical Cartography: Why the Map Is Never the Territory* (London, 2021)

Cooper, David, Christopher Donaldson and Patricia Murrieta-Flores, *Literary Mapping in the Digital Age* (Abingdon, 2016)

Cosgrove, Denis, ed., *Mappings* (London, 1999)

Couldry, Nick, and Anna McCarthy, eds, *MediaSpace: Place, Scale and Culture in a Media Age* (London, 2004)

Crampton, Jeremy, 'Cartography: Performative, Participatory, Political',
     Progress in Human Geography, XXXIII/6 (2009), pp. 840–48
Dando, Christina E., *Women and Cartography in the Progressive Era*
     (Abingdon, 2018)
Dodge, Martin, Rob Kitchin and Chris Perkins, eds, *Rethinking Maps:
     New Frontiers in Cartographic Theory* (Abingdon, 2009)
Dunn, Stuart, *A History of Place in the Digital Age* (Abingdon, 2019)
Edney, Matthew, *Cartography: The Ideal and Its History* (Chicago,
     IL, 2019)
Egbert, Simon, and Matthias Leese, *Criminal Futures: Predictive
     Policing and Everyday Police Work* (Abingdon, 2021)
Fuchs, Christian, *Digital Capitalism: Media, Communication and Society*
     (Abingdon, 2022)
Fujikane, Candace, *Mapping Abundance for a Planetary Future*
     (Durham, NC, 2021)
Gabrys, Jennifer, *Program Earth* (Boston, MA, 2016)
Gay'wu Group of Women, *Songspirals: Sharing Women's Wisdom of
     Country through Songlines* (Sydney, 2020)
Graham, Mark, and Martin Dittus, *Geographies of Digital Exclusion:
     Data and Inequality* (London, 2022)
Haraway, Donna, 'Situated Knowledges: The Science Question in
     Feminism and the Privilege of Partial Perspective', *Feminist Studies*,
     XIV/3 (1988), pp. 575–99
Iosefo, Fetaui, Anne Harris and Stacy Holman Jones, *Wayfinding and
     Critical Autoethnography* (Abingdon, 2021)
Kennedy, Helen, and Martin Engebretsen, *Data Visualization and
     Society* (Amsterdam, 2020)
Kitchin, Rob, Tracey P. Lauriault and Matthew W. Wilson, eds,
     Understanding Spatial Media (London, 2017)
Kurgan, Laura, *Close Up at a Distance: Mapping, Technology, Politics*
     (New York, 2013)
Lindner, Christoph, and Gerard F. Sandoval, *Aesthetics of Gentrification:
     Seductive Spaces and Exclusive Communities in the Neoliberal City*
     (Amsterdam, 2021)
Lordan, Robert, *The Knowledge: Train Your Brain Like a London Cabbie*
     (London, 2018)
McDaniel, John, and Ken Pease, *Predictive Policing and Artificial
     Intelligence* (Abingdon, 2021)
Marres, Noortje, *Digital Sociology: The Reinvention of Social Research*
     (London, 2017)
Massey, Doreen, *For Space* (London, 2005)

Middleton, Jennie, *The Walkable City: Dimensions of Walking and Overlapping Walks of Life* (Abingdon, 2022)

Mindrup, Matthew, *The Architectural Model: Histories of the Miniature and the Prototype, the Exemplar and the Muse* (Boston, MA, 2019)

Monmonier, Mark, ed., *The History of Cartography*, vol. VI: *Cartography in the Twentieth Century* (Chicago, IL, 2015)

Næss, Hans Erik, *A Sociology of the World Rally Championship* (London, 2014)

O'Connor, M. R., *Wayfinding: The Science and Mystery of How Humans Navigate the World* (New York, 2018)

Olsson, Gunnar, *Abysmal: A Critique of Cartographic Reason* (Chicago, IL, 2010)

Pakenham, Thomas, *The Scramble for Africa: 1876–1912* (London, 2003)

Peluso, Nancy Lee, 'Whose Woods Are These? Counter-Mapping Forest Territories in Kalimantan, Indonesia', *Antipode*, XXVII/4 (1995), pp. 383–406

Perkins, Chris, 'Cultures of Map Use', *Cartographic Journal*, XLV/2 (2008), pp. 150–58

Pickles, John, *A History of Spaces: Cartographic Reason, Mapping and the Geo-Coded World* (Abingdon, 2004)

Playful Mapping Collective, ed., *Playful Mapping in the Digital Age* (Amsterdam, 2016)

Powell, Alison, *Undoing Optimization: Civic Action in Smart Cities* (New Haven, CT, and London, 2021)

Reddleman, Claire, *Cartographic Abstraction in Contemporary Art: Seeing with Maps* (Abingdon, 2019)

Rose, Gillian, *Visual Methodologies* (London, 2016)

—, ed., *Seeing the City Digitally: Processing Urban Space and Time* (Amsterdam, 2022)

Rose-Redwood, Ruben, et al., 'Decolonizing the Map: Recentering Indigenous Mappings', *Cartographica*, LV/3 (2020), pp. 151–62

Rossetto, Tania, *Object-Oriented Cartography: Maps as Things* (Abingdon, 2019)

Sheller, Mimi, *Mobility Justice: The Politics of Movement in an Age of Extremes* (New York, 2018)

Skarlatidou, Artemis, and Muki Haklay, *Geographic Citizen Science Design: No One Left Behind* (London, 2021)

Sletto, Bjørn, et al., *Radical Cartographies: Participatory Map Making from Latin America* (Austin, TX, 2020)

Solnit, Rebecca, *Infinite City: A San Francisco Atlas* (Los Angeles, CA, 2010)

Tyner, Judith, *Women in American Cartography: An Invisible Social History*
(London, 2019)

Vaughan, Laura, *Mapping Society: The Spatial Dimensions of Social
Cartography* (London, 2018)

Wilken, Rowman, *Cultural Economies of Locative Media* (Oxford, 2020)

Wilmott, Clancy, *Mobile Mapping: Space, Cartography and the Digital*
(Amsterdam, 2020)

Wilson, Matthew W., *New Lines: Critical GIS and the Trouble of the Map*
(Minneapolis, MN, 2017)

Zukin, Sharon, *The Innovation Complex: Cities, Tech, and the New Economy*
(New York, 2020)

ACKNOWLEDGEMENTS

This book is the culmination of ten years of research on maps and how people use them in everyday life. Its aim is not to provide the definitive guide on maps, but rather to act as a marker of my understanding so far. *All Mapped Out* is a title that could be read in two ways: mapping out my research about maps, or a personal and exhaustive reflection that this is all I have to say about maps at the moment. I'll leave it up to the readers to decide.

Most of my thanks for this book goes to the people whose work I have cited throughout. Citation remains an academic practice not normally given much thought in the wider non-fiction market. Authors are expected to 'own' what they say. I cannot claim to own what I say, when it is clear to me that what I say has been shaped by others. I often doubt whether I have anything original to say at all! For me, citation is more than an act of professional courtesy dictated by the publisher. It is an act of acknowledgement and respect for the ideas of others that inform your own, even if you do not necessarily agree with everything that is being said. And for those like me who *still* write out their references by hand (I know, I know), we are reminded of the people and ideas that came to shape our thinking every time we are forced to format them for publication. So, for details of the works that continue to shape and inspire me, please read the References.

Citational acknowledgements aside, I thank the individuals that have helped me to develop my research and to refine the themes in the book. First, I would like to thank Max Edwards, without whom the idea to write this book would not have come about. Those initial discussions and constructive efforts to hone the proposal during the madness of 2020 helped me to see what this project could become. I also thank the many people who helped in the process of bringing the book to press, most notably Michael Leaman, Amy Salter, Alex Ciobanu and Susannah Jayes at Reaktion Books, as well as Rosie Fairhead and Jo Stimfield for their copy-editing and proofreading, respectively, and my esteemed colleague, Astrid Van den Bossche, whose generous feedback on early drafts was indispensable during the editing phase. Then there are those colleagues and friends that, unbeknown to them, helped steer the book in better directions. Thanks to Sam Hind, Hans Erik Næss, Kaushiki Das, Savyasachi Anju Prabir, Ashwin Matthews, Laura Gibson, Pete Chonka, Kimbal Bumstead, Jina Lee, Sterling Mackinnon and Johnny Henshall for taking the time to listen to me and enlighten me with their expertise, and to Lottie Chesterman for coming up with the title. Special thanks also to Veronica della Dora, Harriet Hawkins, Phil Crang, Oli Mould, Innes Keighren and Pip Thornton of Landscape Surgery fame, whose bi-monthly meetings I am only realizing now were an intellectual luxury in an otherwise commodified industry. To Jenny Harding, Glen Hart and Jeremy Morley for access to and support from the Ordnance Survey, and to the Livingmaps team, where I have found a comfortable home to discuss maps and learn about new mapping practices. Special thanks to Phil Cohen for his continued encouragement, and to my fellow *Livingmaps Review* editors, Barbara Brayshay, Debbie Kent, Clare Qualmann, Blake Morris and William Illsley, who continue to inspire in the ways they bring new work on mapping to my attention.

Finally, thanks to Liz as always, whose continued support is something I should remember not to take for granted. Thank you so much for all that you do.

PHOTO ACKNOWLEDGEMENTS

The author and publishers wish to thank the organizations and individuals listed below for authorizing reproduction of their work.

Illustration by Mona Caron, cartography by Ben Pease, design by Lia Tjandra: p. 121 (this map originally appeared in Rebecca Solnit, *Infinite City: A San Francisco Atlas* (Berkeley and Los Angeles, CA, 2010)); Dong E, Du H, Gardner L. (An interactive web-based dashboard to track COVID-19 in real time): p. 58 (Johns Hopkins Coronavirus Resource Center/*Lancet Inf Dis.*, xx/5, pp. 533–4, https://coronavirus.jhu.edu/map.html); Damien Griffiths: p. 124; courtesy of Johnny Henshall, 2021: p. 65; Andrew Howe: p. 62; Jina Lee: p. 113 top and bottom; Velodyne Lidar: p. 161; Sol Pérez Martínez: p. 61; courtesy of the Mercator Institute for China Studies: p. 54; Pixabay Images: p. 166 top and bottom; Researchgate: p. 60 (Filippo Arrieta/Public Domain/Uploaded by Tom Koch); courtesy of Team O'Neil Rally School: p. 39; 'Vote Leave' Campaign: p. 99; Wikimedia Commons: pp. 6 (Luca Giarelli/cc-by-sa 3.0), 9 (Ruparch/ Footsteps of Man, 1996/cc-by-sa 3.0), 13 (Phil Uhi/cc-by-sa 3.0), 53 (F. von Richthofen, 1877/Public Domain), 59 (John Snow/Originally published in 1854 by C. F. Cheffins, Lith, South-hampton Buildings, London, England/Public Domain), 82 (Elkanah Tisdale/ Public Domain), 96 (Günther Zainer/Public Domain), 106 (Richard de Bello/Public Domain), 108 (Michelin/Public Domain), 134 (Adalbert von Rößler/Public Domain); Jeremy Wood: p. 67.

INDEX

Page numbers in *italics* refer to illustrations